Banister Fletcher

Carpentry and Joinery

A text-book for architects, engineers, surveyors, and craftsmen

Banister Fletcher

Carpentry and Joinery

A text-book for architects, engineers, surveyors, and craftsmen

ISBN/EAN: 9783337384302

Printed in Europe, USA, Canada, Australia, Japan

Cover: Foto ©berggeist007 / pixelio.de

More available books at **www.hansebooks.com**

"The Builder" Student's Series.

CARPENTRY AND JOINERY.

A TEXT-BOOK FOR ARCHITECTS, ENGINEERS,
SURVEYORS, AND CRAFTSMEN.

FULLY ILLUSTRATED

AND WRITTEN BY

BANISTER F. FLETCHER,

Associate of the Royal Institute of British Architects; Vice-President of the Architectural Association; Joint Author, "A History of Architecture"; Lecturer on Architecture and Building Construction, and Director of the Studio, King's Coll., Lond.; Late Examiner in Carpentry and Joinery to the City and Guilds of London Institute,
&c., &c.,

AND

H. PHILLIPS FLETCHER,

Associate of the Royal Institute of British Architects; Fellow of the Surveyors' Institution; Director of the City Companies' Trades Training School, Great Titchfield Street, W.; Lecturer on Quantity Surveying, &c., King's Coll., Lond.; Joint Author, "Fletcher on Quantities,"
&c., &c.

LONDON:
D. FOURDRINIER, 46, CATHERINE STREET, W.C.
1898.

PREFACE.

THE Authors are assured that a concise book, both for reference and for the instruction of students, on the important subjects of Carpentry and Joinery is much needed. That there are very skilful and elaborate works on these subjects nobody could deny; but it is generally admitted that they are cumbersome, sometimes verbose, and at the present time certainly in many ways not in accord with modern practice.

When the craft was threatened, in some of its branches, by the introduction of iron and steel, the more skilful disposition of parts, and the minuter calculation of strains became necessary, and our Trans-Atlantic brethern have developed these essentials to an enormous extent, as those who have visited their country must have noticed.

The Authors have endeavoured to meet the requirements of the craftsman, and at the same time to produce a work that will be useful to the Professional man in the designing of the various structures. They have also endeavoured to consider the desires of those who are likely to become candidates for the examinations of the City and Guilds Institute, the Carpenters' Company, and the

Institute of Certified Carpenters, &c. Also for the examination in these subjects by the R.I.B.A. and the Surveyors' Institution, &c. The Authors have had considerable experience in examining candidates in these subjects, and they feel that some such work would be welcomed by these candidates.

The convenient system of *Tabulation* and *Comparison* has been used where possible, and over 420 illustrations have been specially prepared by the Authors to illustrate their text.

The greater part of the contents of this book originally was written for, and appeared in, the Student's Column of *The Builder.*

BANISTER F. FLETCHER.
H. PHILLIPS FLETCHER.

29, New Bridge Street,
  Ludgate Circus, E.C., Nov., 1897.

CONTENTS.

	PAGE
CHAPTER I.—VARIOUS WOODS IN USE AND THEIR CHARACTERISTICS	9
CHAPTER II.—TIMBER—THE GROWTH, SEASONING, AND CONVERSION INTO SCANTLINGS	14
CHAPTER III.--TOOLS USED IN CARPENTRY AND JOINERY	22
CHAPTER IV.—JOINTS USED IN CARPENTRY ...	30
CHAPTER V.—ROOFS, I.	40
CHAPTER VI.—ROOFS, II.	52
CHAPTER VII.—BRIDGES	63
CHAPTER VIII.—SHORING AND STRUTTING ...	76
CHAPTER IX.—CENTRES	89
CHAPTER X.—SCAFFOLDING, STAGING, AND GANTRIES	99
CHAPTER XI.—PILLARS, BEAMS, AND GIRDERS ...	112
CHAPTER XII.—FLOORS	121
CHAPTER XIII.—FLOOR-COVERINGS	135
CHAPTER XIV.—FRAMING IN PARTITIONS AND FRAME HOUSES	142

CONTENTS.

	PAGE
CHAPTER XV.—ORNAMENTAL CARPENTRY	153
CHAPTER XVI.—JOINTS USED IN JOINERY.— HINGING	165
CHAPTER XVII.—MOULDINGS	176
CHAPTER XVIII.—DOORS	189
CHAPTER XIX.—WINDOWS	201
CHAPTER XX.—FRAMING	213
CHAPTER XXI.—SKYLIGHTS AND LANTERNS ...	225
CHAPTER XXII.—STAIRCASES	236
CHAPTER XXIII.—SHAPED WORK	248
CHAPTER XXIV.—BEVELS	261
CHAPTER XXV.—RODS	273
CHAPTER XXVI.—WORKSHOP PRACTICE	277

CARPENTRY AND JOINERY.

CHAPTER I.

VARIOUS WOODS IN USE AND THEIR CHARACTERISTICS.

THE following are the principal soft woods in general use :—
Northern pine (*Pinus sylvestris*) is commonly known as yellow deal, and it is in general use for all building purposes. The best varieties come from Dantzig, Memel, Riga, and Archangel ; the inferior kinds from Scotland, Sweden, and Norway. The wood is easily worked, and is tough, elastic, and moderately light ; it is used for joists, flooring, roofs, sash-frames, and scaffolding, &c. ; it shrinks from one-twentieth to one-thirtieth in the process of seasoning, and the annual rings should not be more than one-eighth of an inch thick. Inferior kinds have thick rings filled with resinous matter. The weight is about 32 lbs. per cubic foot.

Spruce (*Abies excelsa*) is commonly called white deal ; it is used for cheap buildings where not exposed to the weather, and is much used for the tops of dressers and for shelves, and also for scaffold-poles, and comes from the north of Europe, that from Christiania being the best. It is liable to warp, twist, and shrink, and is knotty, and therefore, hard to work. There are several varieties of American spruce (*Abies alba*) now on the market, though these as a rule are inferior to the above.

American yellow pine (*Pinus strobus*), sometimes known as Weymouth pine, because introduced by Lord Weymouth, is inferior in strength to Baltic timber and very subject to dry rot. It can always be distinguished by the hair-like short streaks running in the direction of the fibres. It is used by joiners and cabinet makers. Its specific gravity is about 35 lbs. per foot cube, or about 3 lbs. more than Northern pine.

B

Canadian red pine (Pinus rubra) is so called from the colour of its bark. It is used for internal fittings, is tough, elastic, clean, and fine in grain, and very durable where air has free access.

Kauri pine (Dammara Australis) is used for joinery, and is the strongest pine known. It is a yellowish-white wood which planes up with a silky lustre and fine close grain. It is used largely for yacht masts. It comes only from New Zealand, and averages about 36 lbs. per cubic foot.

The following are the principal hard woods in use :—

Pitch-pine (Pinus rigida) is used for the best structural purposes; it is an excellent wood for joinery, but it has great specific gravity, and the difficulty in working it entails expense in its use. Pitch-pine comes principally from Virginia. It is liable to cup and heart shake (see next chapter).

Oak is the strongest and most durable wood for ordinary building purposes. It has distinct medullary rays, and shrinks about one-thirtieth of its width in seasoning. Owing to the presence of gallic acid in its composition it causes iron fastenings to corrode. English oak is taken as a standard of quality for all woods, and weighs about 48 lbs. per cubic foot.

Austrian and German oak is very hard and tough, and is largely used, and is specially cultivated in large forests.

American oak is one of the best foreign oaks, being but slightly inferior to English. Riga oak is commonly called wainscot oak (although the latter is a term often used to indicate any oak cut along the medullary rays so as to show the silver grain); it has fine, highly-figured silver grain, is used for panelling and for the very best joinery. "Crown Riga" is the best quality.

African oak is known as the link between teak and oak. It is shipped from Sierra Leone, and is mostly used for ships. It is of a dark reddish colour, free from defects, but difficult to work owing to its close grain. It is one-third stronger than English oak, and weighs about 57 lbs. per cubic foot.

Chestnut was much used for roofs and other carpenter's work in the Middle Ages, but is now not much used except for posts and palings. It resembles oak in appearance, but is slightly darker in colour, and has no medullary rays and no sap-wood. It is easier to work than oak, but not so strong, and liable to give way under a cross strain. The tree itself is one of the largest grown in Europe, and lives to a great age, though when old it is very brittle. Chestnut is imported also from Africa and North America.

Mahogany is generally used for handrailing, ornamental joinery, and cabinet work; as it will not stand the weather it should not be used for external work. That which is shipped from Cuba (called *Spanish mahogany*) is, as a rule, sound and free from defects, but its pores contain a white chalky substance, which dulls the edge of the plane, and makes it harder to work than the other species; it has a fine wavy grain and takes a very high polish. It weighs about 64 lbs. per cubic foot.

Honduras mahogany, from Central America, is a stronger wood than Spanish, and has a transverse strength very nearly equal to British oak. It is largely used for panelling and fascia boards, owing to its immunity from warping. It is not usually attacked by worms or insects, but requires great care, as if seasoned too rapidly it is subject to large shakes. It has a clear straight grain. Inferior qualities may be detected by the appearance of grey specks.

Mexican mahogany has some of the characteristics of Honduras, but is usually obtainable in large sizes; it is considered liable to star shakes, and though it is a good wood for joinery, it is not equal to Honduras or Cuban.

Yarrah (or Jarrah), sometimes called Australian mahogany, is a species of gum tree from Western Australia. It is of a dark red colour, and has a close wavy grain; it is durable and rigid. It is, however, deficient in tenacity; and in seasoning narrow cracks of great depths are sometimes formed. It is well adapted for piers, piles, and other heavy work, and has of late years been much used with some success for paving roads in the form of small longitudinal blocks.

Teak, sometimes called Indian oak, comes from Burmah and Southern India. It is straight in grain, and of a warm brown colour; it somewhat resembles English oak, but has no medullary rays. It is light and easy to work, and possesses an aromatic resinous oil, which not only makes it durable, but enables it to resist the attacks of worms and insects. If the oil has collected into the shakes and hardened, it destroys the tools and is hard to work; this oil, however, preserves iron from rust, and thus teak is much used for backing to armour-plates and in railway contracts. The species which is shipped from Moulmein, in Burmah, is generally considered the best, and weighs about 47 lbs. per cubic foot; that from Malabar, however, is preferred by some.

Greenheart is the strongest wood in use. It comes principally from Demerara and British Guiana. It has a fine, hard, and close grain, is of a dark chestnut colour, is full of minute pores, and the sap-wood is difficult to distinguish from the heart. It is one of the least fire-resisting of all hard woods, and is liable to split at the ends in seasoning. It is much used for piles and marine structures. Greenheart is the heaviest wood in use, and weighs nearly 70 lbs. per cubic foot.

Ash combines elasticity and toughness to a remarkable degree, and should be used in any work that is subject to shock or strain exerted suddenly. It is of a brownish-white colour with yellow streaks in the direction of the grain, with very distinct pores. It has no sap-wood and is used more by the wheelwright than by the carpenter. The variety grown in England is considered the best, though that from America and Canada is largely imported for the manufacture of oars, &c. Good specimens weigh about 50 lbs. per cubic foot.

Elm is twisted in the grain and is of a reddish-brown colour. It is very durable in either wet or dry situations, but lasts but a short time under varying conditions. It is not liable to split, and takes nails well. When Old London Bridge was demolished, the elm piles, which were some hundred years old, were found to be in a sound condition. The sap-wood withstands decay as well as the heart. The best varieties are grown in England, France, and Spain. Dutch elm is very inferior. A variety known as Wych elm is grown in Scotland and the north of England, and is much used for boat-building. Canada also supplies elm of a whitish-brown colour, but the sap-wood is liable to decay. The best elm weighs about 42 lbs. per cubic foot.

Walnut is of a varying brownish tinge. The Italian variety is much prized by the cabinet-maker. English walnut, and especially that shipped from America, is used for window frames and sashes, doors, and the best class of joinery, also for gun-stocks (as it contains no oil that will affect metal), and veneers, but it is too flexible for constructive work. It weighs about 41 to 43 lbs. per cubic foot.

Beech has a hard clean surface and fine even grain, and is not difficult to work. It is used for mallets, planes, piles, and chair-making, and has such a remarkable cleavage that it is used for band-boxes and sword-scabbards. Besides being grown in England, it is found in the temperate parts of Europe, America, and Australasia. It weighs about 45 lbs. per cubic foot.

Amongst other woods seldom used by the carpenter and joiner are *hornbeam*, which has no sap-wood, and is used for turning and mallets; *birch* and *cedar*, which are used principally for furniture; *acacia*, from which good trenails, posts, and railings are made; *poplar*, which is used for light purposes, and stands the weather well; *American plane*, which is hard to work but is durable under water; *larch*, which is well suited for floor-boards and stairs, and also posts, railings, and scaffold-poles, where there is much wear; and, lastly, *alder*, which is occasionally used for turnery and cabinet-work, and is, moreover, extremely durable under water. The buildings of the City of Ravenna are built on piles made of this wood, as are also the abutments of the Rialto Bridge at Venice.

CHAPTER II.

TIMBER.—THE GROWTH, SEASONING, AND CONVERSION INTO SCANTLINGS.

TIMBER grows by successive concentric layers, called "annual rings," so called from one being formed yearly; these annual rings are deposited as *Sapwood* at the circumference, and gradually harden into what is called *Heart-wood*. At right-angles to these concentric rings, and radiating from the centre, are the *Medullary Rays* (fig. 1). The process of formation into heart-wood varies from nine to thirty-five years, according to the nature of the tree. Those trees which perform this hardening in the quickest time are nearly always the most durable. The Sap, in temperate zones, rises in the Spring from the roots of the tree through the cellular tubes to form the leaves, and flows back again, chiefly between the wood and the bark, in the autumn, thus forming the new annual ring.

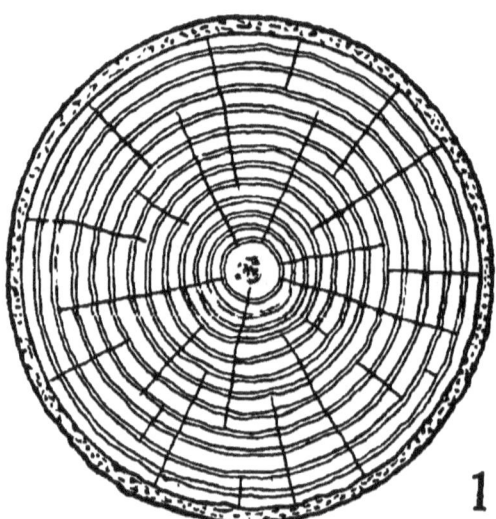

1. Growth of Timber.—Annual rings and medullary rays.

FELLING.—The tree is at its best for felling just before maturity, that is to say before the topmost branches have ceased to be strong and green and to put forth shoots. The heart-wood is stronger and more lasting than the sap-wood, and the former should always be used in good work. The best time for felling oak is when it is about one hundred and fifty years old; for ash, elm, and larch when they are about eighty years old; for fir from seventy to one hundred years old; and poplar at about forty-five years old.

DEFECTS.—Timber is liable to the following defects: *Heart-shake* (fig. 2) is a flaw extending from a cavity, produced by decay, at the heart to the bark; it is usually found in trees which have passed their maturity.

Cup-shake (fig. 2) is a cavity formed between two or more of the annual rings; it sometimes extends a great distance up the trunk of the tree: pitch-pine is very liable to this defect.

Star-shake is a defect somewhat similar to heart-shake, but the clefts radiate from the centre without any appearance of decay; they often render the tree utterly unfit for conversion. If the shakes are large or continuous the timber should be discarded. When the decayed part is white it will probably be found to be not very serious; but if yellow, the timber should not be used.

When the wood is red, or *foxy*, the tree will generally be found to be utterly useless, at all events for constructive purposes.

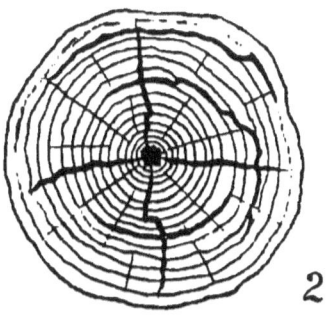

2. Defects of Timber.—Heart-shake and Cup-shake.

Druxiness is a form of decay occasioned by the rain penetrating through a puncture in the bark caused by the tearing off of a branch, as, for instance, in a tempest.

Rindgalls are caused by the growth of timber over a wound caused by a branch being lopped off.

Dry rot is caused by the dissemination of the germs of fungi occasioned in confined places where the gases cannot get away, through want of ventilation. Dry rot occurs only after the tree has been felled.

Wet rot occurs in the growing tree where the timber has been allowed to become saturated, and the communication of this disease is occasioned by actual contact.

GOOD TIMBER.—The qualities of good timber are that it should be straight in fibre, free from large, loose or dead knots, and shakes of every kind. It should be sweet to the smell, the surface should not be woolly or clog the teeth of the saw, but should be firm and bright, and when planed should have a silky lustre. A disagreeable smell betokens decay, and a chalky appearance is generally a sure sign of decomposition. The annual rings should be regular and the colour uniform. Naturally coloured timber (such as mahogany, &c.) should be dark, as this indicates durability and strength. When struck with the hammer good timber should be sonorous, and a gentle tap at one end of a balk should be distinctly heard when the ear is placed at the other. In comparing two pieces of timber of the same species, the heavier is generally the stronger. In fir or pine for the best work, imperfections of any kind should cause the rejection of the wood, but in some of the harder woods a slight defect is generally allowed to pass. Timber of different species should not be used if possible in conjunction with each other; though Alberti, the Florentine architect, may be considered pedantic in suggesting that all timber in one building should be from the same forest. Timber should be felled in winter as the sap is then down in the roots.

SEASONING.—In seasoning, timber should be protected from the weather, but should be allowed a free circulation of air. After drying slowly it is probably better to quarter large trees. Logs are more quickly seasoned by boring a hole down the centre, this also prevents splitting, which may also be stopped by hooping the ends with iron. Speaking generally, *seasoned timber* for carpenters' work should have lost one-fifth of its weight, and *dry timber* for joiners' work about one-third. A squared block of oak takes, approximately, one month for every inch in depth to season, and five times as long to dry. Fir should take a month for every two inches for seasoning, and five times this amount is generally considered sufficient for drying. Timber if covered takes about one-third less time to season than if exposed. *Water seasoning* is carried out by immersing the timber for a fortnight or more in fresh or salt water, and then drying in the air. This process, it is maintained, washes out the sap and renders the timber less liable to warp, but it renders it more brittle, and, in the case of seawater seasoning, causes it to attract moisture after it is fixed. Boiling and steaming quicken the process of seasoning, but are generally considered to reduce the strength and elasticity of the timber. Hot-air seasoning is sometimes used for small scantlings, but it is generally admitted that it reduces the strength. By charring the ends of posts is prevented the attacks of worms and the growth of fungi.

Percentage of Loss of Weight in Seasoning.

Red pine	12—25	Oak	16—25
American yellow pine	18—27	Elm	35—40
Larch	16—25	Mahogany	16—30

ARTIFICIAL SEASONING.—There are several artificial processes of seasoning, some of which have been used with success. These may be divided into two classes:

1. *The sap may be expelled by hydraulic pressure, and replaced by chemical fluids;*

2. *Or the timber may be impregnated with some fluid, which acts on the sap and prevents its decay.*

Of the former, *Bouchere's* process of sulphate of copper and *Blythe's* process of carbolic or tar acids are the best known. Of the latter, the more used are: *Bethel's* process consists of impregnating the fibres with creosote or oil of tar; *Kyanising* or immersing the timber in a saturated solution of corrosive sublimate (bichloride of mercury); and *Burnetising*, which consists of immersion in a solution of chloride of zinc. These methods are mostly dependent on the porosity of the timber for their success, and it may be noted that fluid forced into timber travels one hundred times quicker with the grain than across it.

Conversion of Timber.

To cut the best beam from a log (fig. 3), divide the diameter, AB, into three equal parts, and draw lines at right-angles to

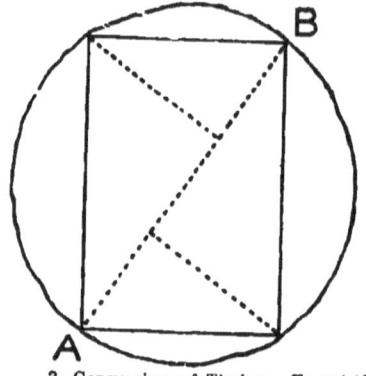
3. Conversion of Timber.—To cut the best beam.

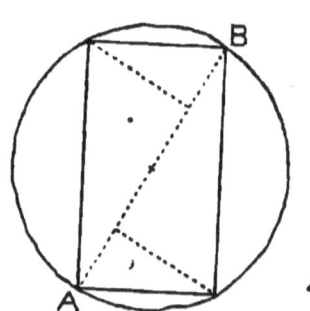
4. Conversion of Timber.—To cut the stiffest beam.

the diameter, from the two points thus obtained, to the circumference, and form the parallelogram by joining the four points on the circumference.

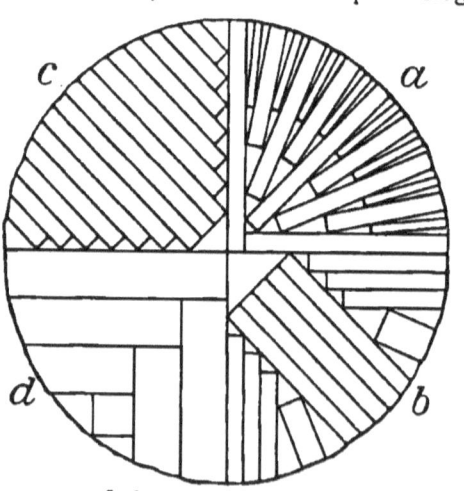
5. Conversion of Timber.—Oak.

To obtain the stiffest beam (fig. 4), divide the diameter, AB, into four equal parts and proceed as before.

In converting oak from the log, care should be taken to have the cuts converging or tending towards the heart, as this will prevent splitting and warping, and the stuff will only then shrink in its width; the best method is shown at *a* (fig. 5).

The medullary rays (or silver grain) are shown to the best advantage by this method, which is very important in joinery, and there is practically no waste, as the triangular pieces are used for feather-edged laths, for fencing, and for tilting pieces. The method shown at *b* in this figure is a good one, if several pieces of the same size are required, as shown together in the centre of this quadrant, though the silver grain will not show to

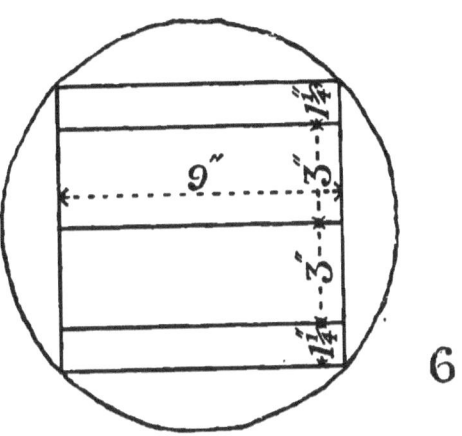

6. Conversion of Timber.—To cut deals from log.

such advantage. The method indicated at *c* does not produce such good timber, and the silver grain will scarcely show at all, while that at *d* will be very liable to warp and twist, and no silver grain will be visible. In obtaining *deals* from the log, consideration must be given to the existing demands of the market; fig. 6 is the method of conversion now generally in use: the two 9 in. by 3 in. deals would come to the London market, and the two 9 in. by 1¼ in. deals would go to the French market, which is the favourite mart for all deals which do not suit the London contracts. The *standards* by which timber is usually sold are as follows:—

Name.	No. of pieces.	Size.			Cubic feet.	Feet super.
		ft.	in.	in.		
Petersburg	120	12 ×	11 ×	1½	165	1,980
London and Dublin ..	120	12 ×	9 ×	3	270	1,080
Christiania	120	11 ×	9 ×	1¼	103½	990
Christiania	60	15 ×	11 ×	1½	103⅜	825
Quebec long hundred	120	10 ×	11 ×	3	275	1,000
Quebec short hundred	100	12 ×	11 ×	2½	229¼	1,100
Drammen	120	9 ×	6½ ×	2½	121⅝	585

There are various marks and brands cut and scribed on timber by shippers, and there are also "quality" marks, which vary according to the port of shipment. The marks shown on figs. 7, 8, and 9 indicate first, second, and third or middling quality, as generally shown on timber coming from Memel, and are placed near the end of the balk. When, however, the timber is shipped from Dantzic, the quality marks consist of a single line, as shown in fig. 7; the first, seconds, and thirds being indicated by the number of short lines or crosses marked across this line. Swedish deals are branded on the ends in red and sometimes in black paint. Norwegian marks are generally in blue paint. American deals, by some shippers, are marked in imitation of Baltic brands

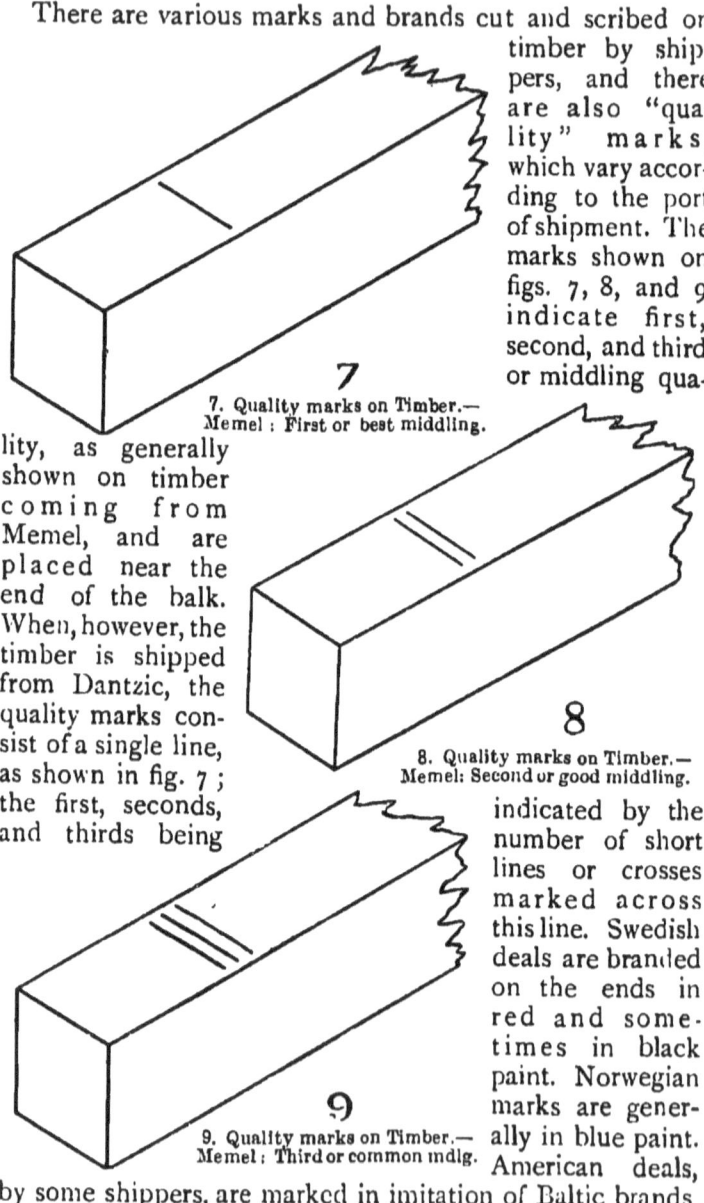

7. Quality marks on Timber.—
Memel : First or best middling.

8. Quality marks on Timber.—
Memel: Second or good middling.

9. Quality marks on Timber.—
Memel: Third or common mdlg.

but the qualities are generally distinguished by the marks I, II, III, in red upon their sides or ends.

For measuring timber, the following is the best formula :—

$$C = L \left(\frac{G + g + g^1}{3} \right)^2$$

G = one-fourth girth of tree at middle in feet.
g = ,, ,, ,, at one end in feet.
g^1 = ,, ,, ,, at other end in feet.
L = length of log in feet.
C = cubic contents of log in feet.

Note.—Deduct thickness of bark from each quarter girth before working out. This varies from half an inch to two inches.

Timber wrought one side loses one-sixteenth of an inch, and if wrought both sides one-eighth of an inch in thickness.

The following definitions may prove of use to the student :—

A *balk* is a roughly-squared log.

A *plank* is from two to six inches thick, eleven inches broad, and generally from 8 ft. to 21 ft. long.

A *deal* is nine inches broad and not more than four inches thick.

A *batten* is similar to a deal, but not more than seven inches broad.

A *square* is 100 ft. super.

A *hundred* contains 120 deals.

A *load* contains 50 cubic ft. of squared timber or 40 of unhewn timber, or 600 superficial feet of inch-planking.

CHAPTER III.

THE VARIOUS TOOLS USED IN CARPENTRY AND JOINERY.

It is intended in this section to give a general outline of the various tools used by the carpenter and joiner, and the purposes for which each is used, with illustrations. A *bench* is the first requisite of the carpenter and joiner, and one of the first tests should be to see that it is strong and firm; the top surface level and even; and that it is not only constructively sound, but that the wood is well seasoned; the height should be about 2 ft. 9 in., and the width not less than 2 ft.; the length being dependent upon the space that is available in the shop; one of its long sides is provided with a vertical side-board with drilled holes to admit of pegs for holding "stuff" to be planed, which is gripped at the other end by a bench-screw; the latter is best if made on the instantaneous principle, that is, if the lever is raised the vice can be simply expanded or contracted by the mere effort of pulling or pushing, and by dropping and turning the lever the bench-screw is rendered rigid.

BENCH-STOPS, against which the wood rests when being planed, are best if made in the form of two wedges, tightening against each other in a mortise cut in the bench-top. At the present time there are many patents on the market in which iron and steel play an important part; there is

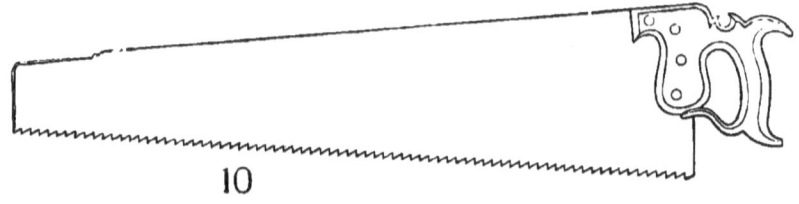

10. Ripping-saw.

often a danger, however, of the tool slipping into the iron and having its edge damaged thereby.

SAWS.—The *saw* is generally the first tool used by the craftsman, and, although so much sawing is now done by machinery, scarcely any work can be carried out without its aid.

The *ripping-saw* (fig. 10) is used for dividing wood in the direction of its fibres, and has generally eight teeth to every three inches in length, the cutting edge of these teeth is at right-angles to the line which ranges with the points; the length of the blade is about twenty-eight inches. The *half-ripper* is about the same length, is used for the same purpose, but has nine teeth to the three inches.

The *hand-saw* for cutting against the fibre is about 24 in. in the blade, and has 15 teeth to 4 in. The cutting edge of these teeth incline forward at angle of 80 deg., or 10 deg. more than for those saws which are used for cutting with the fibre.

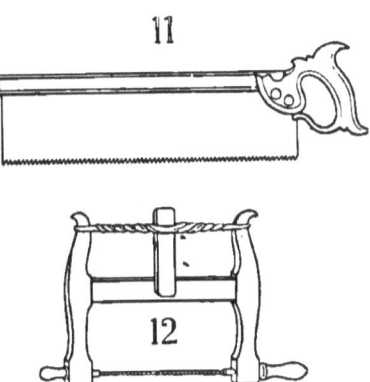

11. Tenon-saw.   12. Frame or Bow Saw.

The *panel-saw*, which has a very thin blade with 6 teeth to the inch, which are very finely set, is used for cutting thin wood. The *tenon-saw* (fig. 11) has a very thin blade of about 14 in. long, 10 teeth to the inch, and is strengthened on the upper edge by a back of sheet-iron; it is used for cutting across the fibres, such as in the shoulders of tenons, &c. The *sash-saw* has an 11-in. blade, with 13 teeth to the inch; it has a brass back to strengthen it, and is used for forming the tenons and mitres of sashes. The *dovetail-saw* has a 9-in. blade, with 15 teeth to the inch; it has a brass back, and is used for the purposes its name implies. The *compass-saw* has a narrow blade, ¼ in. wide at the end and 1 in. at the handle, with about 5 teeth to the inch; it is used for circular work, and is made thicker on the cutting edge. The *keyhole-saw* resembles the compass-saw, but it has a movable handle which can be fastened at any point along the blade; it is used for cutting round quick curves. The *frame (or bow) saw* (fig. 12), by means of which a fine, thin ribbon-saw is placed in severe tension, which can be increased or diminished at will by twisting the slip of wood in the twine, held in position by the cross-bar, is chiefly used for fret work, the blades being easily made to twist round to suit any pattern.

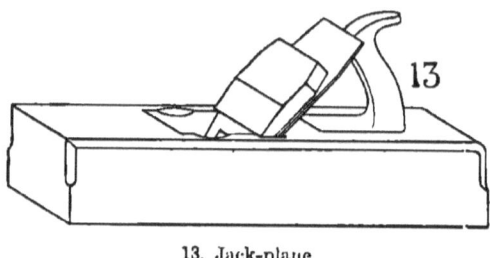

13. Jack-plane.

A saw should have a thin blade, should be dark in colour, have a clear ring when struck, and a "Toledo"-like temper, the point springing more than the heel. On referring to figs. 10, 11, it will be observed that there are two hooked projections in the handle, one above and one below the grasp; if the hand bears on the upper one increased pressure is given to the teeth near the point; if on the lower, the pressure is on those near the handle. The teeth in a saw are generally bent alternately towards the opposite sides of the blades; this is called the *set* of the saw, and is done in order to clear the wood from the teeth.

PLANES.—Among the bench-planes used by the joiner, the *jack-plane* (fig. 13) is used to take off the rough surface

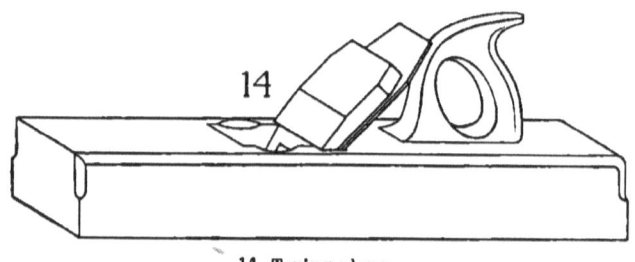

14. Trying-plane.

of the wood; and it is usually about 15 in. long in the stock and 3 in. wide. The cutting-edge of the iron is convex, for the purpose of separating the surface fibres, and this produces a surface with slight irregular corrugations. The next process is to change these undulating surfaces to one level surface; this is accomplished by the *trying-plane* (fig. 14), which is 18 in. to 24. long, 3½ in. wide, and has an iron with a straight edge. In working, this plane is not stopped at the limit of the arm's length, as when using the jack-plane, but is made to take a shaving off the whole length of the "stuff." The mouth of the trying-plane is much narrower than that of the jack-plane, and consequently takes off a much finer shaving, and by its greater length tends to correct any deviation from a perfectly plane surface, consequent upon the depth of bite of the the jack-plane. After the "trying-up" has been performed, it will often be found that, owing to irregularities in the fibres of the wood, some parts, although level, are left rough; this is rectified by the use of the *smoothing-plane* (fig. 15), which is short, being about 8 in. in length and 3 in. wide, has a straight cutting edge, and is also used for cleaning off finished work. A *'ointer* is 30 in. long, and is used for "shooting," which means planing up the edges of boards perfectly straight, to form a close joint with adjoining boards. The foregoing comprise all those technically known as the *bench-planes*.

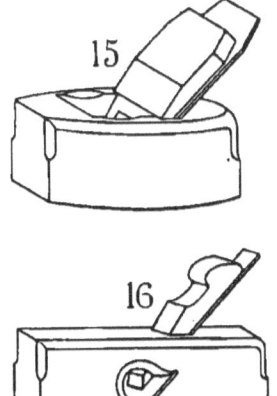

15. Smoothing-plane. 16. Rebate-plane.

The *compass-plane* is similar to a smoothing-plane, except that its "sole" or under side is convex, and is thus used for forming a concave surface. The *forkstaff-plane* is similarly constructed, but its sole is concave, and it is thus used for obtaining convex surfaces. The *rebate-plane* (fig. 16) is different to those previously described, in that the iron reaches to the edges of the stock in its width so as to enable the cutting edge to be carried into the corner of the rebate, for the sinking of which it is used as its name implies; it is usually 9 in. long, and varies in width from ½ to 2 in. The

c

sash-fillister is a rebate-plane for sinking the edge of the stuff that is away from the craftsman, and the *moving-fillister* or *side fillister* for sinking the edge next to the workman. The *plough-plane* (fig. 17), of which an improved form is shown, is used for forming grooves at varying distances from the edge of the stuff, the distance being regulated by the screw. *Match-planes* are a species of plough-plane used for cutting (*a*) the groove and (*b*) the tongues in what is called match boarding, the iron of each is shown (fig. 18). The *bead-plane* is used for sticking a moulding whose section is semicircular. The *snipebill-plane* is used for forming the quirk on either side of the bead; and the *router or old woman's tooth* (No. 19) is used for clearing out grooves across the grain as in staircase strings. There are also many forms of *moulding-* or *fillister-planes*, which it is not deemed necessary to mention, as most of this work is now performed by rotating cutters worked by steam-power.

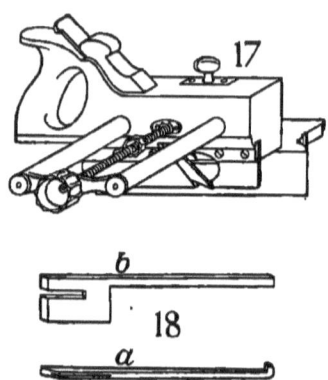

17. Plough-plane. 18. Match-plane Irons.

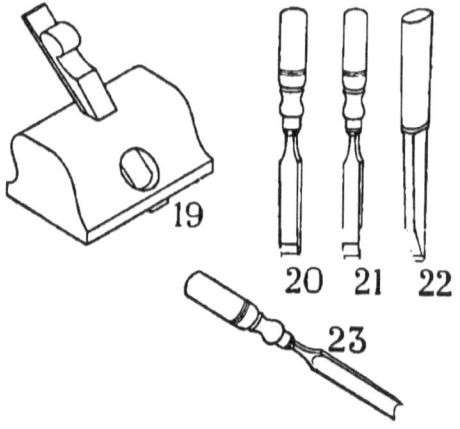

19. Router. 20. Firmer-chisel. 21. Paring-chisel. 22. Mortise-chisel. 23. Gouge.

CHISELS. — The various forms of *chisels* may be divided into the following classes:—The *firmer-chisel* (fig. 20) has a steel face, and is used for heavy work with the mallet; the *paring-chisel* (fig. 21) is of slighter make, has a very fine edge, and is used with a thrusting motion of the hand or shoulder for taking off thin shavings; the *mortise-chisel* (fig. 22) is used for cutting mortises by the aid of a mallet. The *gouge* (fig. 23) is a convex iron used for cutting concave surfaces; the *drawing-knife* has its edge set at an angle with its length, and is used for drawing in the ends of tenons so as to guide the course of the saw.

The *spokeshave* is practically a two-handed chisel, and is used by drawing towards the craftsman, by which the depth of the cut is governed by the depth of the blade from the haft. It is used for curved surfaces and edges, and is much in request by wheelwrights.

The *brace and bit* (fig. 24) admits of that uninterrupted rotation of a tool which cannot be obtained by the use of

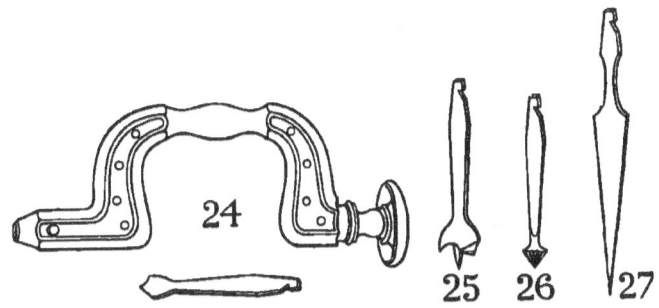

24. Brace and Bit. 25. Centre-bit. 26. Counter-sink. 27. Taper-bit.

the awl, gimlet, or auger. The *bits* used are various in character to answer different purposes: the principal are the *centre-bit* (fig. 25), for boring large holes; the *counter-sink* (fig. 26), for enlarging a hole to admit of the head of a screw or bolt being placed flush with the surface of the stuff; and the *taper-bit* (fig. 27), for forming funnel-shaped holes.

The *square* (fig. 28) generally consists of a wooden stock or tack with a steel blade fitted into it exactly at right-angles and secured by three rivets or screws; a square is indispensable, it is employed for trying-up the surface of the stuff and for setting out joints, &c.

C 2

CARPENTRY AND JOINERY.

Bevels (fig. 29) differ from squares in that they are used for marking lines at angles other than a right angle, the metal blade being movable on a screw joint; care must be taken to screw it up tightly or errors will be the result.

There are three kinds of *gauges* in general use. The *marking-gauge* (fig. 30) consists of a sharply-pointed spike driven into a shank about nine inches long with a sliding block which can be fixed at any point along its length by means of a wooden screw; this gauge is much used for marking parallel distances by being run along shot edges of the stuff. The *cutting-gauge* is similar to that just described, but has a thin steel plate instead of a spike, sufficiently sharpened as to be capable of making a cut

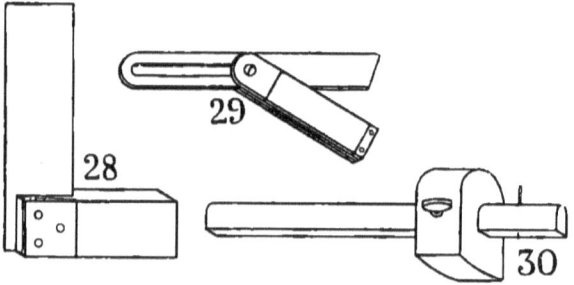

28. Square. 29. Bevel. 30. Marking-gauge.

with or against the grain; the *mortise-gauge* is similar in principle to the others, with the addition of a second adjustable spike, which enables the two parallel lines of a mortise or tenon to be traced on the stuff at one operation.

The *side-hook* or *sawing-rest* consists of a strip of wood 9 in. long, with a small block at each end on opposite sides; in use, one end hangs over the bench and against the other the piece of stuff is thrust; the sawing-rest is useful for cutting or planing against the grain.

The *mitre box* is an arrangement for guiding a saw cut at an angle of 45 deg., in its simplest form it consists of two boards at right-angles to each other, attached to a third or bottom board.

The *mitre-square* is an immovable bevel set at an angle of 45 deg. The *straight-edge* is, as its name implies, a slip of wood or iron with a perfectly straight edge, used for marking off straight lines on the stuff.

Two slips or *winding-sticks* of the same dimension, each having two parallel straight edges, are used in order to determine the level plane of the whole surface of a piece of stuff in this manner: after placing them at each end of the surface of the wood, the craftsman, by looking in a longitudinal direction over the upper edges of the two slips, can determine whether they are in the same plane; if they are not, the face of the stuff is *in winding*, if they are, the work is *out of winding*, and satisfactory.

The *screw-driver, hammer, mallet, gimlet,* and *bradawl* are too well known to need description.

CHAPTER IV.

JOINTS USED IN CARPENTRY.

THE methods adopted for joining pieces of wood together have always exercised the ingenuity and skill of the carpenter. In fact, the art of properly and efficiently connecting the various pieces in an assemblage of timbers, lies at the very root of the craft. The subject lends itself very readily to tabulation, for it is evident that joining timbers is necessary under the following circumstances :—

(*a*) For lengthening beams in tension, compression, and cross-strain.

(*b*) Effecting a junction between joists, plates, &c., resting on or in beams.

(*c*) Between upright posts resting on beams.

(*d*) Between beams resting on upright posts.

(*e*) Between oblique struts and beams and struts with posts.

(*f*) For connecting suspending pieces and tie-beams.

Many other joints for different purposes arise, having to suit special circumstances, but they are mostly to be found in combinations of the above-mentioned, and will readily suggest themselves to the student. In connexion with joints, the question of fastenings can hardly be considered apart; these may be briefly summarised as belonging to either of the following types :—Wedges, keys, pins (including wooden pins, nails, screws, and bolts), straps, and sockets. We shall refer to these briefly in the joints in which they are used.

Taking our first division (*a*) *the lengthening of beams* in tension, compression, and cross-strain, we find that three methods are in use, viz., *lapping*, *fishing*, and *scarfing;* these may be described very briefly, as the introduction of iron and steel renders such an operation less usual than formerly. *Lapping* (fig. 31) consists, as its name implies, in laying one beam over the other, and is the simplest form of junction, as illustration; it evidently is of use only for cross-strain or compression, the two pieces being bound together by iron straps, but if a tensile strain is required, bolts passing through the two pieces are used. *Fishing* consists in butting the ends of two pieces of timber together and placing an iron plate of a certain length on each side of the joint and passing bolts through from one side of the beam to the other. A fished joint is unsuitable for a cross-strain, and if used in compression there should be plates on all four sides; the strain is taken by the bolts, and may be lessened in two ways, either by tailing the fish-plates into the beams or by inserting keys, which thus hold

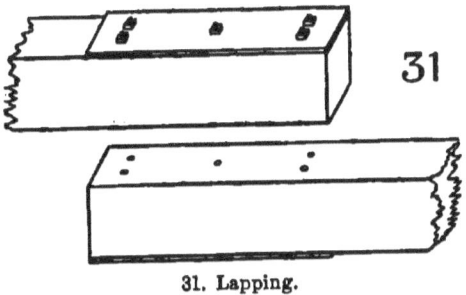

31. Lapping.

the parts firmly together when the strain is put upon them.

Scarfing is a form of jointing in which the surfaces of each piece overlap and are sunk into one another, so that the resulting wooden beams present a much neater finish than either of the foregoing methods. Although the necessity for using such a joint has been minimised by the increased use of iron and steel joists, yet in roofs it is frequently required, as also in country places, where iron is not so easily obtainable. Tredgold goes elaborately into the subject, remarking, however, that the simplest forms are the best in order to ensure accuracy of fitting.

We give the following rules for determining the length of

the different parts of a scarf, according to the qualities of the different varieties of wood in which it is formed.

Proportion of length of scarf to depth of beam, according to Tredgold.

	Without Bolts.	With Bolts.	With Bolts and indents.
Hard wood, as oak, ash, elm	6	3	2
Pine and fir	12	6	4

The sum of the depth of the indents should equal ⅓ the depth of beam.

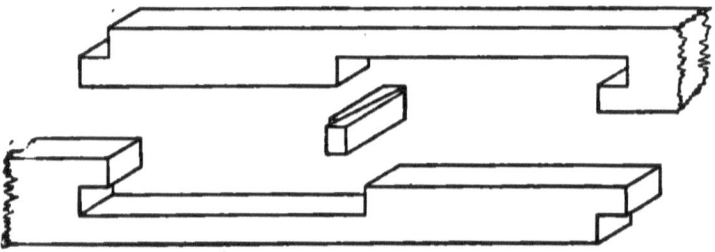

32. Scarfing.—Tension and Compression.

A *scarf to resist tension* is shown in fig. 32, and can be used without straps or bolts. A scarf to resist compression is shown in fig. 33; the principal point to keep in mind is that the bearing surfaces are perpendicular to the strain; the fish-plates tend to keep the joints from buckling or turning over. It is evident that if this form were used for tension, its strength would depend entirely on the bolts, while, if it were used for cross-strain, the latter would tend to bend the iron plates and tear out the bolts.

A *scarf to resist cross-strain* (fig. 34) has the upper fibres in compression and the lower fibres in tension; this being the case, the indents on the upper surface should be perpendicular to the pressure, those in

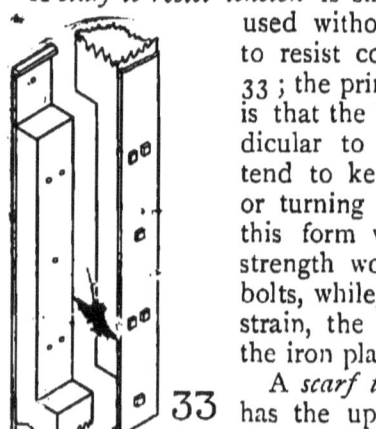

33. Scarfing.—Compression.

the lower portion being oblique to resist tension only, as great a thickness as possible being obtained at *c b*. Barlow mentions that a joint formed in vertical planes between the two connecting-pieces is stronger than the one shown ; it is, however, seldom used.

We have now briefly discussed the joints which may be used in lengthening beams for tension, compression, and cross-strain; it sometimes happens, however, that we may want a joint which can withstand two, or even three, kinds of strain.

A scarf to resist *tension and compression* is shown in fig. 32 ; in this example the resistance to tension is given

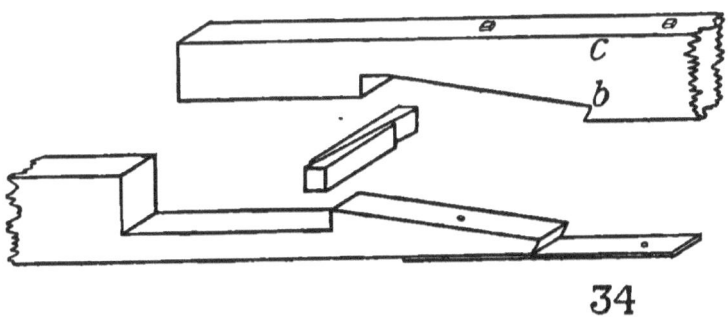

34. Scarfing.—Cross-strain and Tension.

by means of keys of hard wood, or better still, of wedges which bring the pieces close together so that as little strain as possible may fall on the bolts. Keys are usually made one-third the depth of the timber.

A scarf to resist *tension and cross-strain* is shown in fig. 34, the necessary strength for the former being obtained by indenting and the insertion of wedges. Scarfing wall-plates is effected, as shown in fig. 36, but the joint is made in the direction of its length.

(*b*) *Joints for beams bearing on or in beams.*—The simplest form consists in halving (fig. 35), and is generally used for wall-plates ; this joint is effected by sawing half the thickness out of each piece, and, the two being fitted together, the upper and lower surfaces are flush.

Bevelled halving (fig. 36) has the surface of the joint splayed in a bevel form, and a weight being applied in the form, say, of rafters, the two wall-plates are held firmly together.

Dovetail halving is more elaborate still, and its name explains itself. This is not a good joint for carpentry, as wood shrinks considerably more across the grain than along

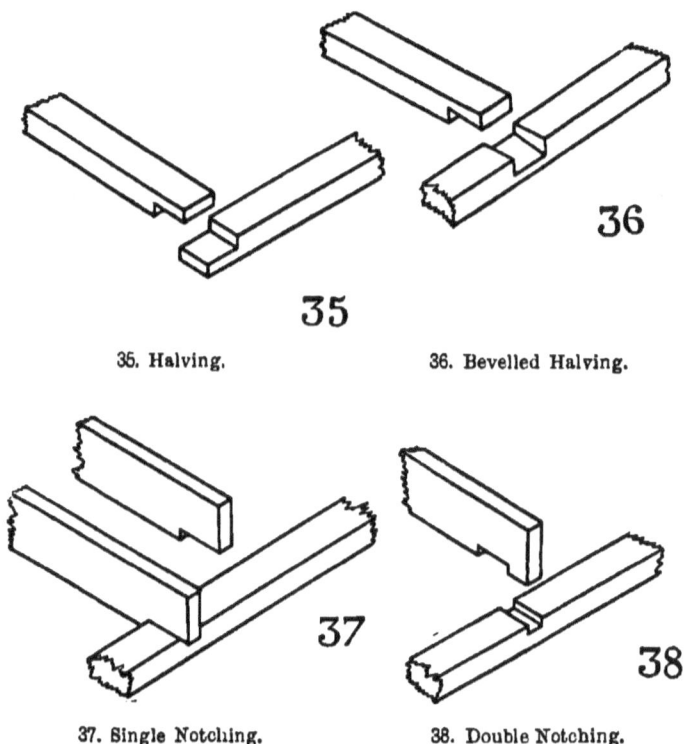

35. Halving.   36. Bevelled Halving.

37. Single Notching.   38. Double Notching.

the length, and consequently the pieces do not fit closely. It is principally used for joining the collar to the rafter in a collar-beam roof.

Single notching (fig. 37) is effected simply by taking a piece out of the lower side of a joist which is to rest on a beam or wall-plate.

Double notching (fig. 38) is so called when two notches are taken out, one from the upper and one from the lower beam; this joint is used where beams are required to be within a certain depth, and saving of space is effected.

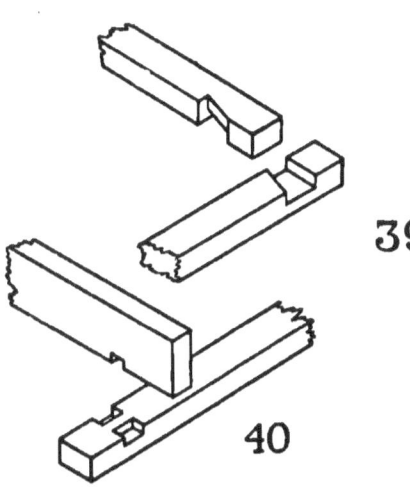

39. Dovetail Notch.   40. Cogging.

Dovetail notching (fig. 39) is principally used for jointing wall-plates at the angles. It is executed as sketch, one of the sides being left straight, and a wedge is sometimes driven in to tighten up the whole.

Cogging (fig. 40) is a form of joint which possesses advantages over notching. The sketch will explain the joint, in which it will be seen that a notch is *partly* cut out in the lower beam, leaving the centre uncut; the upper beam has a notch only sufficiently wide to receive the cog; the advantage of this is that the upper beam is its full thickness at the point of support, and is therefore evidently stronger than when notched. By using this joint for a tie-beam resting on a wall-plate the former thus holds the latter in position, even when it does not project beyond it. Cogging is also used for joint between a purlin and the principal rafter.

Mortise-and-tenon joints (fig. 41).—This joint, in its commonest form, consists of a mortise or rectangular hole cut into the side of the main beam to receive a tenon or projection from the end of the cross-beam, which fits closely into it.

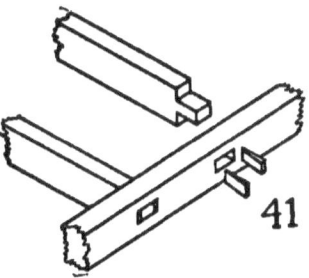

41. Mortise-and-tenon Joint.

Two points to be remembered in designing this important joint are (1) to make the tenon as large and, therefore, as strong as possible, and (2) to make the mortise as small as possible, in order to weaken the beam as little as may be. The mortise and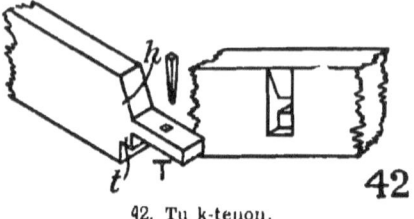

42. Tu k-tenon.

tenon should be placed on the neutral axis, where the cutting of the fibres will weaken the girder least, and where both the mortise and tenon which has to fit into it will be free from tension and compression. In practice the lower edge of the mortise is usually placed on the neutral axis, and the depth of the tenon is $\frac{1}{6}$th of the depth of the beam.

A form of joint called the *Tusk-tenon* is used (fig. 42) in order to give the tenon as deep a bearing as possible, without weakening the beam on which it rests, by largely increasing the mortise. This is effected by adding below the tenon T, the tusk t, which should penetrate the girder $\frac{1}{6}$th of the depth of the joist; above the tenon is formed the *horn h*, which projects the same distance as the tusk. To tighten up the whole joint in a girder of 4-in. thickness, the tenon should be made to project through the girder and pinned by means of pieces of hard wood torn from the

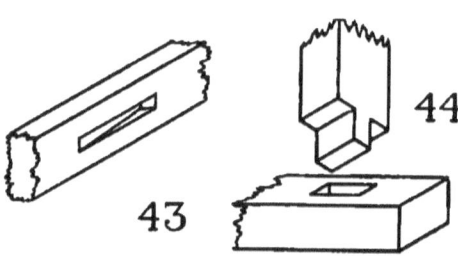

43. Chase-mortise.   44. Upright Mortise.

balk. In thicker girders the tenon should penetrate at least twice its depth, and be pinned through the top of the girder. The trimmer of a hearth is generally tusk-tenoned into the trimming joists.

Chase mortises (fig. 43) are a form of joint used in the

case of two fixed beams between which it is desired to frame a connecting-piece, an oblique chase is cut by means of which this cross-piece is slid along to its place. *Fox-wedging* is used when a beam has to be framed into one already fixed against a wall, the beam, when driven home, is thus made secure.

(*c*) *Joints between upright posts resting on beams* are effected by an ordinary mortise-and-tenon joint, as sketch (fig. 44), in which the width of the upright beam is divided into three, the central division forming the tenon, which is cut shorter than the mortise so that the shoulders of the tenon bear firmly on the sill. The *bridle* joint (fig. 45) is recommended by Tredgold as being easily fitted, and consists of an adaptation of mortise and tenon. The bridle should not exceed one-fifth of the beam in order not to weaken the cheeks of the posts which fit on each side of it.

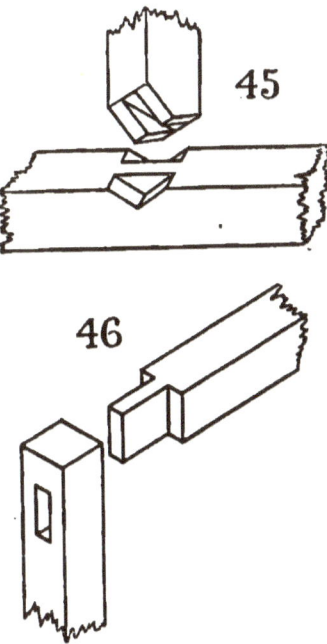

45. Bridle Joint. 46. Post-and-beam Joint.

(*d*) *Joints between beams resting on upright posts* are effected by a horizontal tenon (fig. 46).

(*e*) *Joints between oblique struts and beams and struts with posts.*—The former, as used in the junction of principal rafter and tie-beam is effected by means of a bridle joint (fig. 47). The joint is generally assisted by an iron strap which helps to take the thrust of the rafter and prevent the shearing of the tie-beam. The connexions of principal rafter with king-post (fig. 48) and also the struts with principal rafter are effected by an ordinary mortise-and-tenon joint, as shown at fig. 50. Where the collar of a

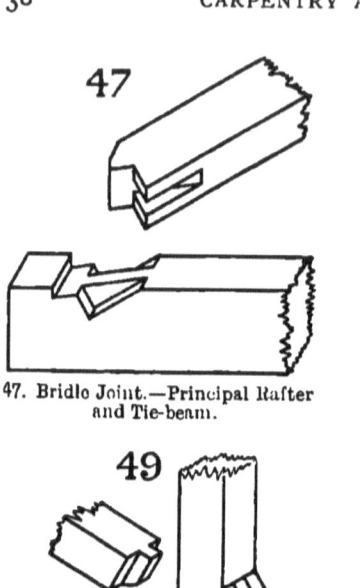

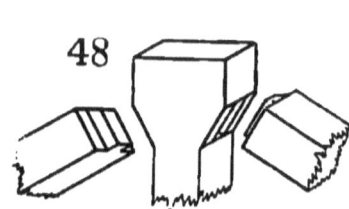

47. Bridle Joint.—Principal Rafter and Tie-beam.

48. King-post and Principal Rafters.

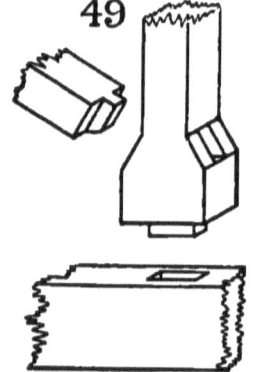

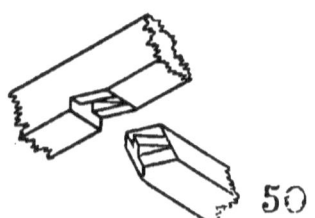

49. King-post and Tie-beam.

50. Strut and Principal Rafter.

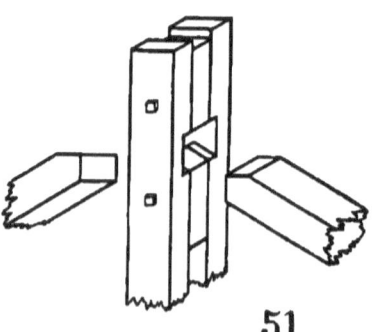

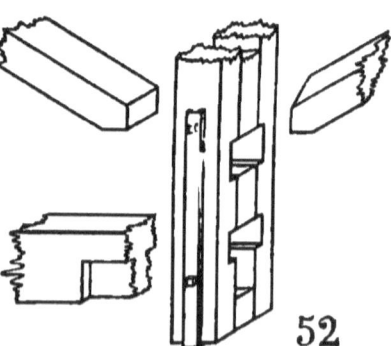

51. Suspending Pieces, Kinghead, and Rafters.

52. Suspending Pieces.—King-post, Struts, Beam and Tie.

roof joins a rafter a dovetail notch is sometimes used, somewhat similar to fig. 39.

(*f*) *The joint between suspending pieces* (*king or queen posts*) *and tie-beams* is effected by means of a mortise-and-tenon joint (fig. 49), which can be braced up by means of what is known as a *gib-and-cottar* joint, which will be described under roofs. A more economical way of making a suspending piece is shown in figs. 51 and 52, in which it is formed of two thicknesses; the rafters (fig. 51) and struts (fig. 52) butt against each other respectively, and the suspending pieces are notched to take up the principal rafters and the tie-beam (fig. 52). Besides the fastenings already mentioned, nails, spikes (*i.e.*, large nails used for heavy work), trenails (or pieces of hard wood), screws, bolts, and various forms of iron straps and shoes are used, which will be referred to later on.

CHAPTER V.

ROOFS.—I.

ON introducing the important subject of roofs to the student, a few general remarks as to the principles to be adhered to in their construction may be advisable, for, as Ware in his "Body of Architecture" has said, "There is no article in the whole compass of the architect's employment that is more important or more worthy of a distinct consideration than the Roof." Without going into any great depth, it is necessary to take a short glance at the strains exerted by roof timbers, in order to enable us to understand the reason and necessity for the employment of such timbers, in connexion with which the special forms bear a direct relation to the strains exerted. It is evident that flat roofs which exert no outward thrust on the walls do not require trussing in any way; in fact, their construction becomes very similar to a floor of similar span, on which "firring" is placed to give the necessary slope to the zinc or lead with which it is covered, to carry off the rain-water.

In a sloping roof, the inclination of its sides to the horizon, which is called the pitch, is regulated by the roofing material with which it is covered, and the country in which it is situated. Many investigators have gone very deeply into this subject, and a writer in the "Encyclopédie Méthodique" has divided the climates of the world into belts or bands parallel to the Equator, each of which requires a certain pitch regulated by the different roof-coverings employed, and he gives a table showing how the roofs of the great buildings of the world in each belt accord with this rule.

Tredgold, the great authority on the subject, gives the

following angles of roofs for different coverings, to which the height of the roof in parts of its span is added :—

Kind of Covering.	Inclination to Horizon.		Height of roof in parts of span.
	°	′	
Copper	3	50	1/30
Lead	3	50	1/30
Zinc	4	0	1/27
Slates, large	22	0	1/5
,, ordinary	26	33	1/4
Asphalted felt	3	50	1/30
Thin slabs of stone or flags	29	41	2/7
Pantiles	24	0	2/9
Thatch of straw	45	0	1/2
Plain tiles	45	0	1/2

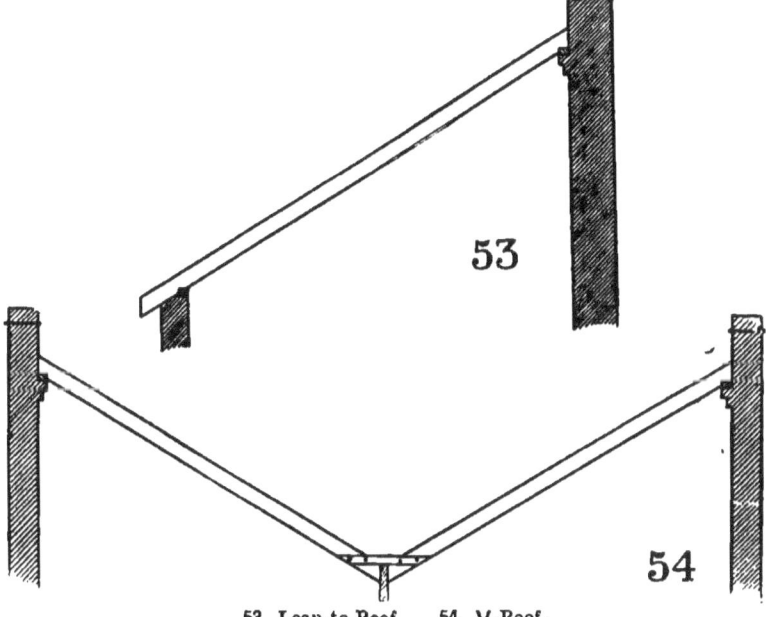

53. Lean-to Roof.   54. V-Roof.

The simplest form of roof is what is known as the *lean-to roof* (fig. 53), in which its one side leans against a vertical wall.

D

A V-*roof* is that in which two slopes incline from side-walls to a central gutter running the length of the building (fig. 54).

A *couple or single-span roof* (fig. 55) is one of the simplest

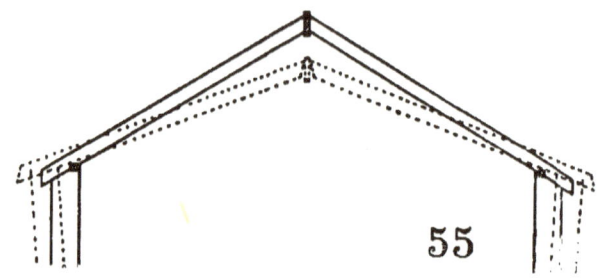

55. Couple Roof.

forms, and consists of two rafters meeting at an inclination upon a ridge-board, to which they are securely nailed; their feet or lower ends being simply notched or nailed on to a wooden plate resting on the top of the wall. It is evident that the weight of the roof-rafters tends to push out the enclosing walls, as indicated by the dotted lines; to remedy this weakness the *couple-close roof* (fig. 56) is adopted. In this form it will be observed that the feet of the rafters are secured from spreading by means of a tie, which effectually

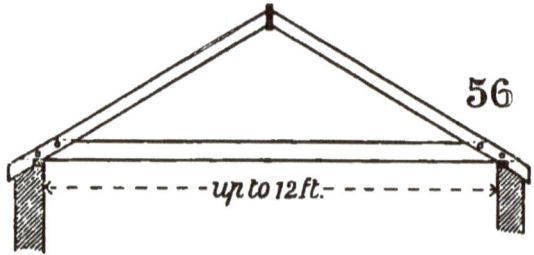

56. Couple-close Roof.

prevents any tendency these may have to press the wall outwards.

The tie-beam, however, especially if it supports a ceiling, is liable to sag in the centre, and in order to prevent this a *king-bolt roof* is employed (fig. 57), a light iron rod being susspended from the ridge and bolted through the tie-beam.

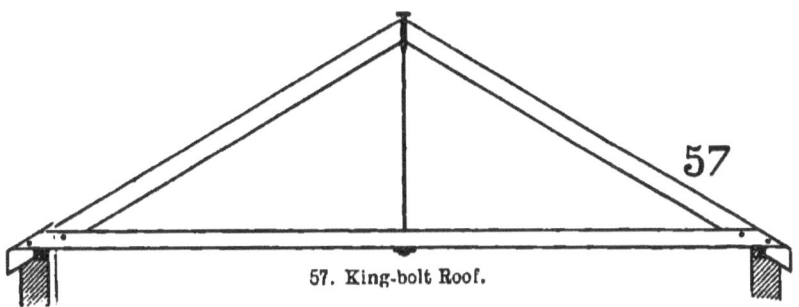

57. King-bolt Roof.

A llight wooden suspending piece, spiked to the ridge, and, at its feet, to a longitudinal piece secured to the centre of thee tie-beams, effects the same object.

Scantlings for Couple-close Roofs.

Pitch 30° with Countess Slates on 1-in. boards.

Span.	Rafter.	Ridge Board.	*Ceiling Joist.
Feet.	In. In.	In. In.	In. In.
8	3 by 2	7 by $1\frac{1}{2}$	4 by 2
10	$3\frac{1}{2}$ by 2	7 by $1\frac{1}{2}$	5 by 2
12	4 by 2	7 by $1\frac{1}{2}$	6 by 2
14	$4\frac{1}{2}$ by 2	7 by $1\frac{1}{2}$	7 by 2
16	5 by 2	8 by $1\frac{1}{2}$	8 by 2
18	$5\frac{1}{2}$ by 2	8 by $1\frac{1}{2}$	9 by 2

* When used they act as ties, but without king-bolt.

Another form of roof for small spans is that known as the *collar-beam roof* (fig. 58). It is useful when part of the roof is to be included in the room. The collar is placed a certain way up the roof (but the nearer to the feet of the rafters the better), and should be spiked or notched on to each rafter (fig. 59). It is intended to form a strut, but if

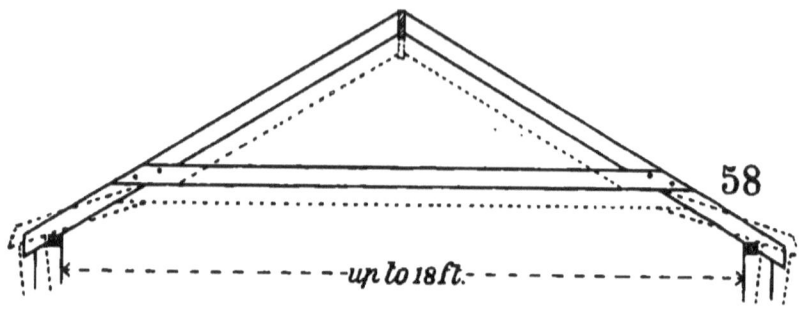

58. Collar-beam Roof.

the walls give way it becomes a tie, and the tendency is for the rafters to take the form shown in the dotted lines. Provided the walls are strong enough, and the collar applied to

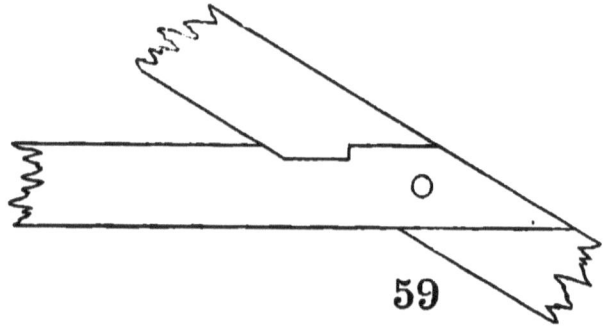

59. Detail of Tension Joint.

each rafter, it makes a fairly good roof; but all the weakness of the construction is emphasised, when, as is often the case, the collars are only placed at intervals, and support a purlin, which, in its turn, supports the intermediate rafters.

The following table is given from the R.E. aide-mémoire for the scantlings used in collar-beam roofs :—

Scantlings for Collar-beam Roofs.

Pitch 30°. If 45° add 1 in. to depth of rafters.

Span in Feet.	RAFTERS. For Countess slates on ¾-in. boards, 1 ft. clear apart, or centre to centre if very exposed site.				COLLARS.	
	Thrust taken by walls or ties, collars ⅓ way up.*		Tied by collars ¼ way up to prevent walls spreading.†		From ¼ to ⅓ way up.	At any height.
	No ceiling.	Ceiled to collars.	No ceiling.	Ceiled to Collars.	No ceiling.	Ceiled to collars.
	Inches.	Inches.	Inches.	Inches.	Inches.	
8	1½ by 2½	1¾ by 3	2¼ by 3¼	2¼ by 3¾	1¾ by 2½	2 in. wide by ½ in. added to every ½ in. for every foot run of clear length of underside of collar.
10	1¾ by 2½	1¾ by 3	2¼ by 4	2¼ by 4½	2 by 2½	
12	1¾ by 2¾	1¾ by 3¼	2¼ by 4½	2¼ by 5	2 by 2¾	
14	1¾ by 3	1¾ by 3½	2½ by 5	2¼ by 5½	2 by 3	
16	2 by 3¼	2 by 3¾	2½ by 5½	2½ by 6	2 by 3½	
18	2 by 3½	2 by 4¼	2½ by 6	2½ by 6½	2 by 4	

* Halve these collars on to rafters without cutting into latter.

† These rafters allow of the tension joint (fig. 59). If the collar is required ⅓ way up, about ¼ in. must be added to both breadth and depth of rafters, and ¾ in. to depth of collars. But with unstable walls, ties are far cheaper, and may be at long intervals, if secured to wall-plates, of sufficient width to take the thrust between the ties.

During the Middle Ages many forms of roofs were used in which the tie-beam at the feet of the rafters was not employed. These will be shortly discussed later on.

We have now arrived at a form of roof which can be used up to 18 ft. Beyond this point, however, it is found that the rafters of ordinary section have a tendency to bend, and require to be supported in their length, and that the tie-beam also must be supported. In order to effect these objects, what is known as "trusses" are used, and spaced at intervals of about 10 ft. apart. The simplest form is known as the *king-bolt truss*, and can be used for spans from 20 ft. to 30 ft. in length. The illustration (fig. 60) shows that in this simple form we have all the elements of a good truss; a wooden tie-beam sometimes supporting a ceiling prevents the feet of the principal rafters from spreading, and is supported in its centre by means of an iron king-bolt which prevents any tendency to sag in the centre. The principal rafters in a span of 20 ft., or more, also require to be supported in the centre of their length. This is effected by means of struts resting at their base against a straining piece, and connected to the principal rafter as shown in fig. 60. The heads of the principal rafters and the king-bolt are let into an iron socket (S, fig. 60), and the feet of the principal rafters are held in position, not only by the oblique mortise and tenon, but also by an iron strap, as shown in fig. 60. The *king-post truss* is constructed on exactly the same principles as the last-named truss, with the exception that, as its name implies, the king-post is of wood. This form of truss is the most important assemblage of timbers which the carpenter produces, and is used in any position where it can be advantageously employed, such as in the upper part of a Mansard roof, and many other cases. This being so, we may well discuss shortly the different parts of which the truss is composed. Fig. 62 is an isometrical view. The trusses are usually set up at distances of 10 ft. (or as near that length as can be arranged), and each truss has, therefore, to be strong enough to bear the weight of two half bays of its length. Across these principal rafters are laid the purlins, upon which rest the common rafters which support the boarding (or battens) on which

CARPENTRY AND JOINERY. 47

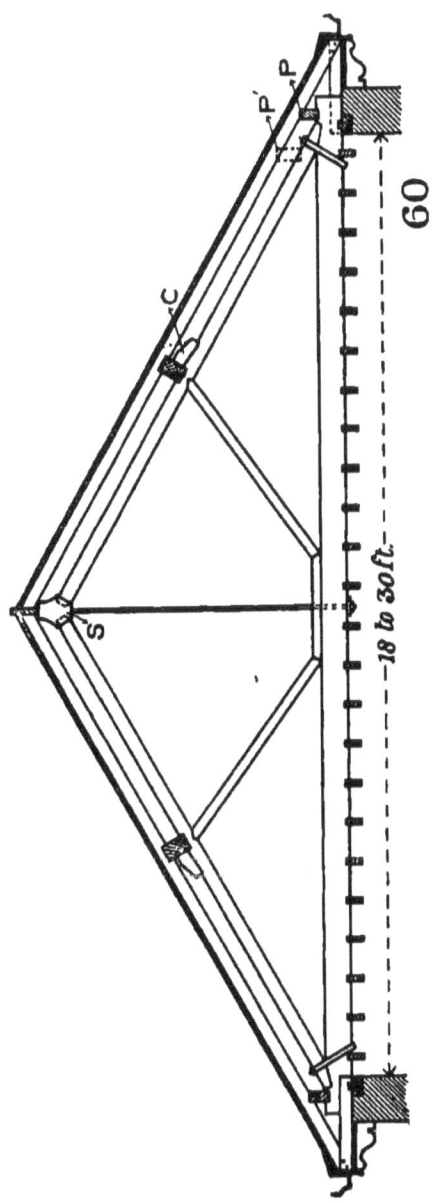

60. King-bolt Truss.

the roof covering is laid. The king-post is primarily a tie to hold up the centre of the tie-beams, and it is further bevelled and mortised at top and bottom to receive respectively the heads of the principal rafters and the feet of the struts (figs. 48, 49), which are tenoned into it. A groove is cut in the head to receive the ridge-piece to which the common rafters are nailed. The king-post is usually further connected at the head with the principal rafter by an iron strap, and to the tie-beam by a stub-tenon and what is known as a gib-and-cottar joint (fig. 61), in order that the tie-beam may be tightened up when the roof is framed up; in this respect care should be taken that the foot of the king-post should, in fixing, be kept well above the tie-beam, in order that it may not bear upon it, instead of supporting it, when the roof settles. *Suspending pieces* (figs. 51, 52) are sometimes used as king-posts, and are more economical and quite as efficient. The method of putting them together is sufficiently shown in the illustration. The *tie-beam*, which holds in the feet of the principal rafters, is subjected to a tensile strain, which is counteracted by the weight of the ceiling, which it occasionally supports. The principal rafters are connected with it by an oblique tenon (fig. 47), and also secured by straps or bolts; the tie-beam is secured to the wall-plate by being notched or cogged, or sometimes rests on stone templates. The *struts*, whose purpose is to support the principal rafter and purlin, are in compression; it is evident they should be placed immediately under the purlins, and it is as well to keep this in mind, although in practice it is sometimes difficult of application; the feet of the struts are mortised into the spreading part of the king-post. The *principal rafters*, connected with the tie-beam and king-post and supported by the struts, are in compression. They support the purlins, which are notched to fit into a cog on the back of the principal rafter. This cog should be as wide as possible, so as not to weaken the principal rafter more than necessary.

The *purlins* are sometimes additionally supported by cleats (C, fig. 60), either spiked or housed to the back of the principal rafter. They should be placed so as to sup-

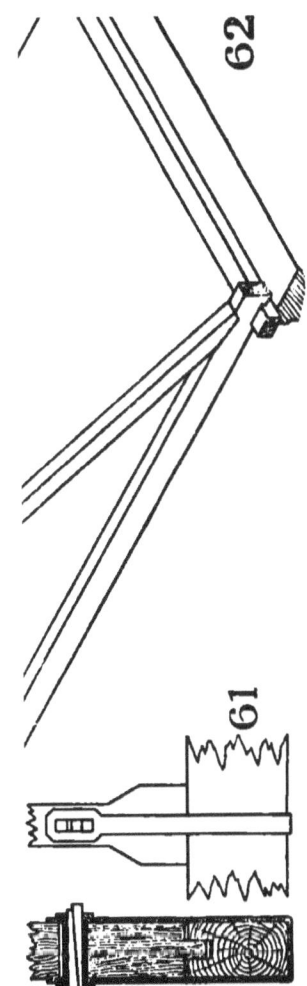

61. Detail of Gib and Cottar.

62. King-post Truss.

Trusses 10 ft. apart.	Span in feet.	Tie-beam, depth includes 3 in. for joints.	Pr ncipal Rafters.	King-posts.	Struts.	Purlins.	Common Rafters.
		Inches.	Inches.	Inches.	Inches.	Inches.	Inches.
King-truss. No ceiling.	20	3 by 4½	3 by 5	3 by 2¾	3 by 3	5 by 7½	2 by 3½
	22	3 by 4¾	3 by 5½	3 by 2¾	3 by 3½	5 by 7¾	2 by 3¾
	24	3½ by 4½	3½ by 5¼	3½ by 2¾	3½ by 3½	5 by 8	2 by 4
	26	3½ by 4¾	3½ by 5¾	3½ by 2¾	3½ by 4	5 by 8¼	2 by 4¼
	28	4 by 4½	4 by 5½	4 by 2¾	4 by 4	5 by 8½	2 by 4½
	30	4 by 4¾	4 by 6	4 by 2¾	4 by 4½	5 by 8¾	2 by 4¾
King-post Truss. Ceiled to Tie-beam.	20	4 by 7	4 by 4	4 by 3	4 by 2½	5 by 7½	2 by 3½
	22	4 by 7½	4 by 4½	4 by 3	4 by 3	5 by 7¾	2 by 3¾
	24	4½ by 8	4¼ by 4¾	4¼ by 3	4¼ by 3	5 by 8	2 by 4
	26	4½ by 8½	4½ by 5	4½ by 3	4½ by 3	5 by 8¼	2 by 4¼
	28	4½ by 9	4½ by 5½	4½ by 3	4½ by 3½	5 by 8½	2 by 4½
	30	4¾ by 9	4¾ by 5¾	4¾ by 3	4¾ by 3¾	5 by 8¾	2 by 4¾

port the common rafter in the middle of their length, and should be formed of as long pieces of stuff as possible, in order to stiffen the whole roof, and to hold the trusses in position and prevent any inclination to heel over sideways.

Pole-plates (P, fig. 60), which run parallel to the length of the roof, are required to take the end of the common rafters, which are bird's-mouthed to them; they should be notched and spiked to the ends of the tie-beam as near the centre of the wall as may be. An objectionable position for the pole-plate is shown by the dotted line (P^1, fig. 60), as it weakens the end of the principal rafter, where it mostly requires strength.

The *common rafters* are supported in their length by the purlin to which they are notched, as also to the pole-plate to which they are also spiked. On these common rafters, the rough boarding is nailed to receive the slates. Authorities seem to agree that by laying these diagonally the roof is strengthened.

We have now touched upon the various joints and connexions in a king-post truss, and from this chapter and that on joints, which has immediately preceded it, the student should have obtained a fair grasp of the subject in order to enable him to understand the construction of roofs of larger span and of more complicated construction, which we shall refer to in our next article.

The preceding table gives the scantlings for king-post roofs with or without ceilings, adopted by the Royal Engineers in the War Department buildings.

The horizontal wind force allowed for is 45 lbs. per foot acting only on one side of roof at a time, or a normal pressure of 30 lbs. per foot for a 30 deg., and 40 lbs. for a 45 deg. pitch.

The scantlings are for pitches up to 30 deg. :—timber, Northern pine; slates, Countess, on 1-in. boards. For roofs of 45 deg. pitch, add 1 in. to the depth of the common rafters, purlins, and struts, and $\frac{1}{2}$ in. to the depth of the principal rafters, as given in table.

CHAPTER VI.

ROOFS.—II.

THE king-post roof-truss may be used up to spans of 30 ft., but as it has been found that the points supporting the tie-beam should not be more than 14 ft. to 16 ft. apart, additional vertical ties known as *queen-posts* are introduced, and the roof shown in fig. 63 is the result, where it will be seen that the tie-beam is supported in two places, and the length of the common rafters is also supported by means of two purlins, one resting on the back of the principal rafter and the other on the straining-beam. In this construction a straining-beam supported by cleats is introduced between the heads of the queen-posts, to hold them in position, the tendency being for the principal rafters to press their heads

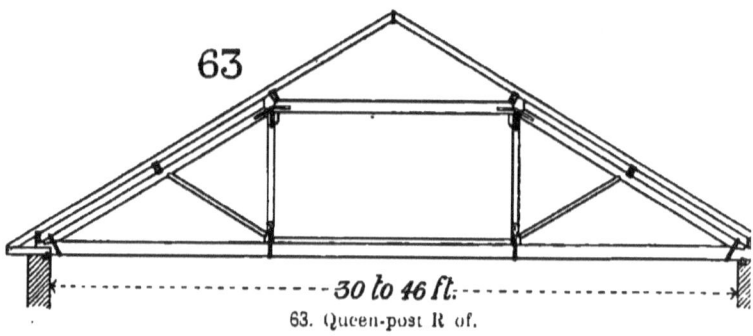

63. Queen-post Roof.

towards each other. On the tie-beam is placed a straining-sill to keep their feet in position, in addition to the stub-tenon with which they are connected to the tie-beam. The queen-post roof-truss is used for spans from 30 ft. to 46 ft. The gib-and-cottar joint is used to the feet of the queen-post for the same purpose as in the king-post truss described in the last article. It will be seen that the queen-posts carry approximately two-thirds the weight of the tie-beam

f any). In roofs over 46 ft. and up to 60 ft.
)eam requires to be upheld in more than two
dditional posts known as *princesses* are intro-
)ending pieces (fig. 64). The straining-beams
50 ft. require supporting in the middle of their
iis is effected as shown, by means of a small
down from the principal rafter continued to
by means of struts from the queen-posts.
er, of this large span are usually in these days
1 iron or steel, and little good can come from
:ir construction in wood, although Tredgold
us examples.
elow the scantlings adopted for queen-post

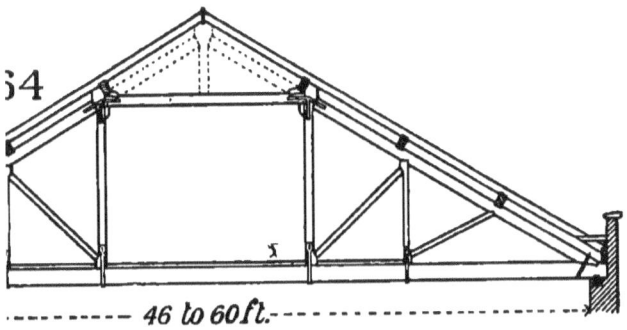

64. Queen-post Roof, with Princesses.

without ceilings, by the Royal Engineers in
artment buildings, as they are considerably
ical than those recommended by Tredgold.
nd force allowed is 45 lbs. per foot, acting
iide of a roof at a time, or a normal pressure
foot for a 30 deg., and 40 lbs. for a 45 deg.
scantlings are for pitches up to 30 deg.
hern pine; Countess slates on 1 in. boards.
45 deg. pitch, add 1 in. to the depth of
rs, purlins, and struts, and ½in. to the depth of
rafters as given in table on next page.

ROOFS.—We have now discussed roofs which
:d in gables, and in which ordinary principals

Trusses 10ft. apart.	Span in feet.	Tie-beam; depth includes 3 in. for joints.	Principal rafters.	Queen-posts.	Struts.	Straining beam.	Purlins.	Common rafters.
		Inches.	Inches.	Inches.	Inches.	Inches.	Inches.	Inches.
Queen-truss. No ceiling.	32	$4\frac{1}{2}$ by $4\frac{1}{2}$	$4\frac{1}{2}$ by $4\frac{3}{4}$	$4\frac{1}{2}$ by $2\frac{3}{4}$	$4\frac{1}{2}$ by $2\frac{1}{2}$	$4\frac{1}{2}$ by $5\frac{1}{2}$	5 by $7\frac{1}{2}$	2 by $3\frac{1}{2}$
	34	$4\frac{1}{2}$ by $4\frac{3}{4}$	$4\frac{1}{2}$ by 5	$4\frac{1}{2}$ by $2\frac{3}{4}$	$4\frac{1}{2}$ by $2\frac{3}{4}$	$4\frac{1}{2}$ by 6	5 by $7\frac{3}{4}$	2 by $3\frac{3}{4}$
	36	$4\frac{3}{4}$ by $4\frac{3}{4}$	$4\frac{3}{4}$ by 5	$4\frac{3}{4}$ by $2\frac{3}{4}$	$4\frac{3}{4}$ by 3	$4\frac{3}{4}$ by $6\frac{1}{4}$	5 by 8	2 by 4
	38	$4\frac{3}{4}$ by $4\frac{3}{4}$	$4\frac{3}{4}$ by $5\frac{1}{4}$	$4\frac{3}{4}$ by $2\frac{3}{4}$	$4\frac{3}{4}$ by $3\frac{1}{4}$	$4\frac{3}{4}$ by $6\frac{3}{4}$	5 by 8	2 by 4
	40	$4\frac{3}{4}$ by 5	$4\frac{3}{4}$ by $5\frac{1}{2}$	$4\frac{3}{4}$ by $2\frac{3}{4}$	$4\frac{3}{4}$ by $3\frac{1}{4}$	$4\frac{3}{4}$ by $7\frac{1}{4}$	5 by $8\frac{1}{4}$	2 by $4\frac{1}{4}$
	42	5 by 5	5 by $5\frac{1}{2}$	5 by $2\frac{3}{4}$	5 by $3\frac{1}{4}$	5 by $7\frac{1}{2}$	5 by $8\frac{1}{2}$	2 by $4\frac{1}{2}$
	44	5 by $5\frac{1}{4}$	5 by $5\frac{3}{4}$	5 by $2\frac{3}{4}$	5 by $3\frac{1}{2}$	5 by 8	5 by $8\frac{1}{2}$	2 by $4\frac{3}{4}$
	46	$5\frac{1}{4}$ by $5\frac{1}{4}$	$5\frac{1}{4}$ by $5\frac{3}{4}$	$5\frac{1}{4}$ by $2\frac{3}{4}$	$5\frac{1}{4}$ by $3\frac{1}{2}$	$5\frac{1}{4}$ by $8\frac{1}{4}$	5 by $8\frac{3}{4}$	2 by 5
Queen-truss. Ceiled to tie-beam.	32	$4\frac{3}{4}$ by $7\frac{1}{4}$	$4\frac{3}{4}$ by $5\frac{1}{2}$	$4\frac{3}{4}$ by 3	$4\frac{3}{4}$ by $2\frac{1}{2}$	$4\frac{3}{4}$ by $6\frac{3}{4}$	5 by $7\frac{3}{4}$	2 by $3\frac{1}{2}$
	34	$4\frac{3}{4}$ by $7\frac{3}{4}$	$4\frac{3}{4}$ by $5\frac{3}{4}$	$4\frac{3}{4}$ by 3	$4\frac{3}{4}$ by $2\frac{3}{4}$	$4\frac{3}{4}$ by $7\frac{1}{2}$	5 by $7\frac{3}{4}$	2 by $3\frac{3}{4}$
	36	$4\frac{3}{4}$ by $8\frac{1}{4}$	$4\frac{3}{4}$ by $6\frac{1}{4}$	$4\frac{3}{4}$ by 3	$4\frac{3}{4}$ by 3	$4\frac{3}{4}$ by $8\frac{1}{4}$	5 by 8	2 by 4
	38	5 by $8\frac{1}{2}$	5 by 6	5 by 3	5 by 3	5 by $8\frac{1}{4}$	5 by 8	2 by 4
	40	5 by 9	5 by $6\frac{1}{2}$	5 by $3\frac{1}{2}$	5 by $3\frac{1}{4}$	5 by 9	5 by $8\frac{1}{4}$	2 by $4\frac{1}{4}$
	42	$5\frac{1}{2}$ by 9	$5\frac{1}{2}$ by $6\frac{1}{4}$	$5\frac{1}{2}$ by $3\frac{1}{2}$	$5\frac{1}{2}$ by 3	$5\frac{1}{2}$ by 9	5 by $8\frac{1}{2}$	2 by $4\frac{1}{2}$
	44	$5\frac{1}{2}$ by $9\frac{1}{2}$	$5\frac{1}{2}$ by $6\frac{3}{4}$	$5\frac{1}{2}$ by $3\frac{1}{2}$	$5\frac{1}{2}$ by $3\frac{1}{4}$	$5\frac{1}{2}$ by $9\frac{1}{2}$	5 by $8\frac{1}{2}$	2 by $4\frac{3}{4}$
	46	$5\frac{1}{2}$ by 10	$5\frac{1}{2}$ by $7\frac{1}{4}$	$5\frac{1}{2}$ by 4	$5\frac{1}{2}$ by $3\frac{1}{2}$	$5\frac{1}{2}$ by 10	5 by $8\frac{3}{4}$	2 by 5

an be used, but cases often arise in which two roofs cut into one another at right angles, or terminate in what is known as a *hip* (figs. 65, 66). In such cases *hip-rafters* (*h*, figs. 65, 66) and *valley-rafters* of deep and narrow section are brought into use, and the jack-rafters are spiked to these. The tendency of the hipped rafter to thrust the wall outwards at angles is counteracted by what is known as a *dragon-beam* D, figs. 67, 68). In this construction the hip-rafter at its base is tenoned into a mortise in the dragon-beam, which is notched on to the wall-plate and supported on its inner edge by being tusk-tenoned into an angle-brace A, which is secured to the wall-plates. In roofs where king and queen-post trusses are employed, and where a hipped-roof is constructed, half-trusses are used at right angles to the main truss to which they are secured.

The question of trimming voids for chimneys, &c., is considered in a later article.

EAVES OF ROOFS, &c.—There are still some points in connexion with roofs which we must briefly consider, as overhanging eaves, dormer windows, and gutters between adjoining roofs and behind parapets. The eaves are the lower portion of the rafters of a roof which overhang the wall and help to protect it from the rain. To the ends of these rafters is fixed an iron gutter to collect the rain-water, which is taken by pipes to the drains. Fig. 69 shows a method of fixing a moulded cast-iron gutter to the feet of the rafters, and fig. 70 a method of securing a half-round gutter by wrought hanging irons. Dormer windows are worked vertically in the side or inclined plane of the roof, and are formed by means of valley-rafters in which the jack-rafters are framed. Dormers are largely used for the lighting of rooms in the roof, and in Mediæval structures have, of course, been very elaborately treated.

GUTTERS—boarded and lead-lined—are either placed behind brick or stone parapets, behind chimneys in a sloping roof, or occur at the meeting of two roofs. The latter are called V-gutters, and are formed, as shown (fig. 71), by framing gutter-bearers between the feet of the rafters near their meeting. V-gutters should be laid to a fall of not less than 2 in. in 10 ft., and they should be not

56 CARPENTRY AND JOINERY.

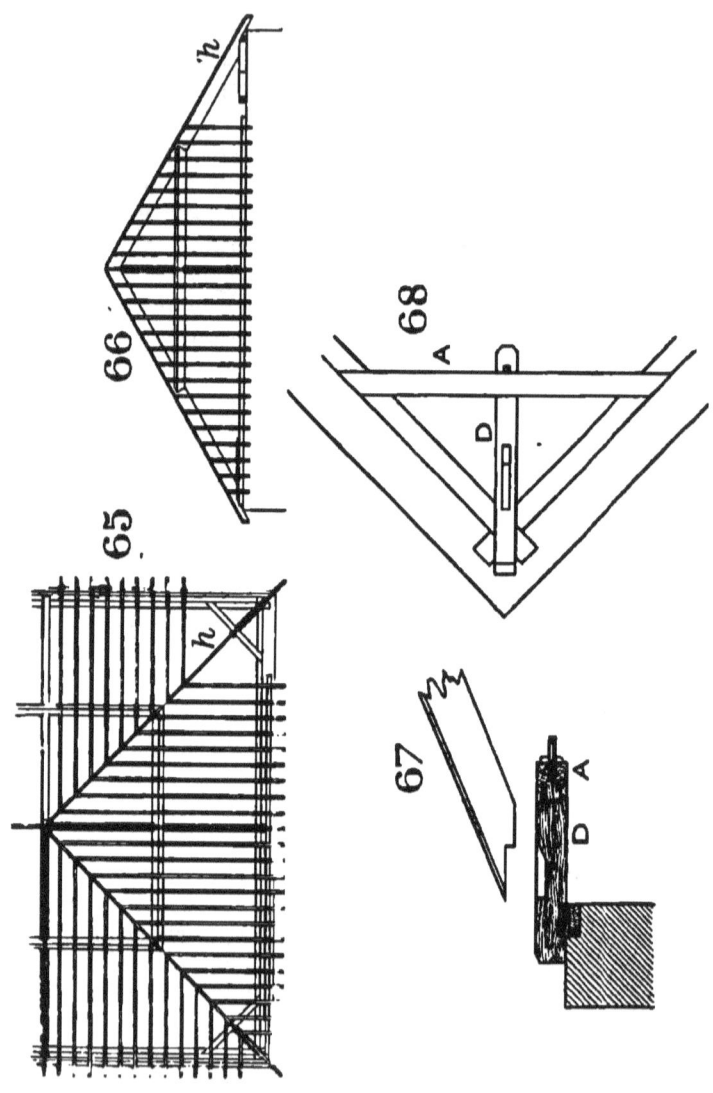

:ss than 9 in. wide at the lowest point. It will be seen
1at these forms of gutters increase in width as they be-
ome higher, because they are framed in the slope of the
ɔof. *Drips* (fig. 72) should be formed every 10 ft. or less
ɔ allow the plumbers to joint the sheets of lead. The
ainwater is collected at certain points into *cesspools*, formed
ʋith dovetailed angles ; these are usually 12 in. square, and
bout 9 in. deep. The rainwater is carried from these
esspools to the rain-water pipe. Lead-flats are formed
imilarly to gutters, and need not be further described here.

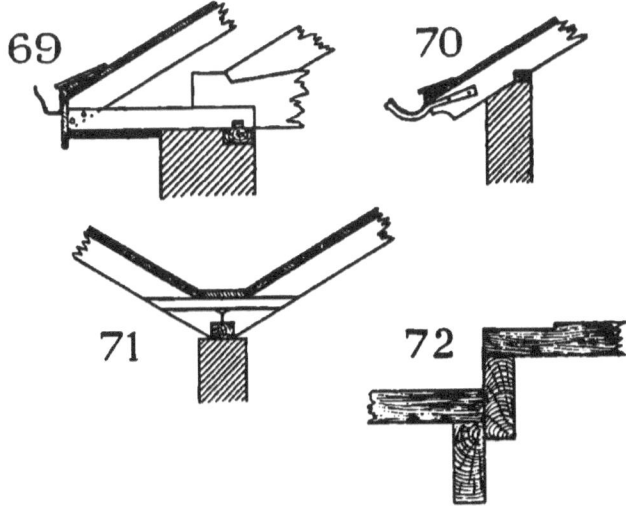

69. Moulded Cast-iron Eaves-gutter.
70. Eaves-gutter supported by Hanging Irons.
71. V-gutter.
72. Drip, with rebate to receive end of lead.

Gutters behind parapets are sometimes formed as V-gutters
r as *box-gutters* ; one of the former form is shown in fig. 73.
'hey are regulated in regard to fall in the same manner as
/-gutters, but being framed of vertical pieces of stuff, are
ιe same width throughout.

MEDIÆVAL ROOFS.—During the Middle Ages, forms of
ɔofs of various types were used. These were either what
; known as " open timber " roofs, or roofs which merely
overed the stone vaulting, and were not visible from below.

E

The combinations of the latter are so various, in order to accommodate themselves to the height of the vaulting, that we can hardly discuss them here; suffice it to say that the principles enunciated in the foregoing examples were applied with ease to these new needs.

In the open timber roofs, the tendency was to do away with the tie-beam at the base of the roof in order to give an appearance of more height to the interior. The pitch adopted was considerably steeper than the slate roofs already noticed; for during the thirteenth century it was

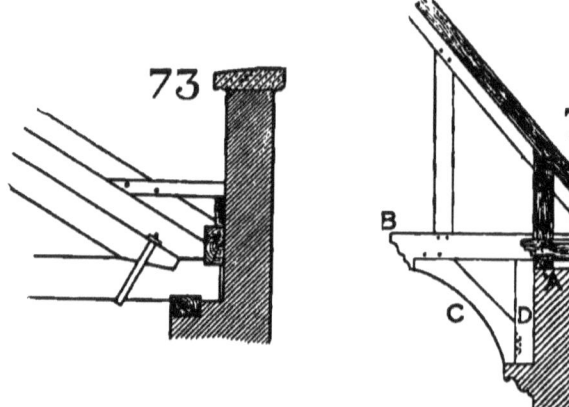

73. Box-gutter.   74. Hammer-beam Roof.

often as much as 60 deg. In the later periods the pitch gradually decreased until at the end of the fifteenth century it became almost flat.

Passing to the open *Hammer-beam* type of roof, which, especially in England, is the great glory of the style, a section is given (fig. 74) from which the principles of the construction can be understood. The hammer-beam BB^1 occupies part of the position of a tie-beam, part of which may have been cut away in order to test the new form. The hatched portion shows the foot of a rafter in a wide wall; the light portion, the position and form of the hammer-beam, the outer end of which is tenoned or halved, and pinned on to the wall-plates A, which were usually placed with

CARPENTRY AND JOINERY.

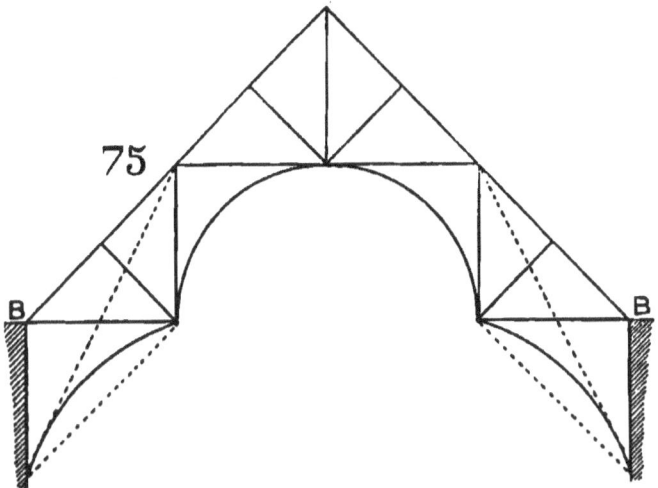

75. Diagram of Hammer-beam Truss.

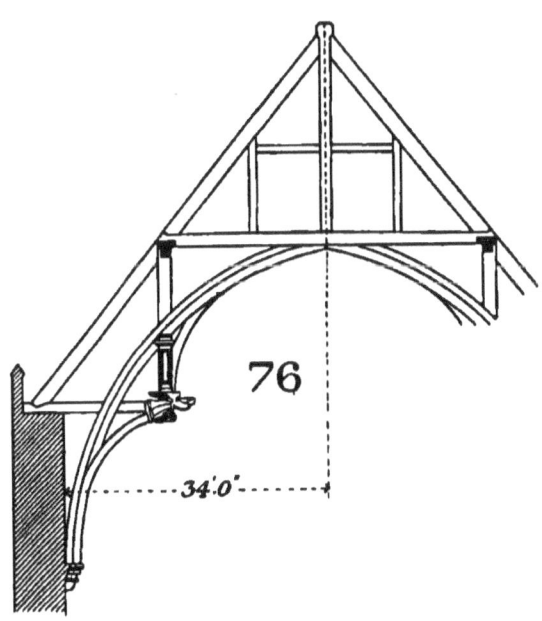

76. Westminster Hall Roof.

their widest sides on to the wall, and were strutted between so as to be thoroughly connected one with the other. It will be observed that the inner end of the hammer-beam is supported by a curved brace C, which is supported on the wall-piece D, which, in its turn, is often supported on a carved corbel stone. The curved brace is mortised, tenoned, and pinned into the wall-piece and inner end of the hammer-

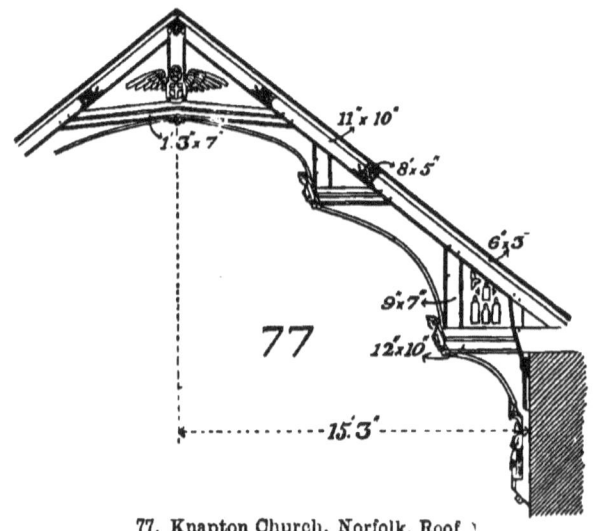

77. Knapton Church, Norfolk, Roof.

beam, which from this point supports a strut E, which is carried up to the principal rafter which it helps to carry.

The importance of this form of construction is that it really shortens the effective span of roof, because the whole length of the hammer-beam may be said to have a solid bearing. Another point to be observed in these roofs lies in the fact that the thrust of the roof, if any, is brough very much lower down the wall which in consequence is less liable to be thrust out. Fig. 75 is a section of a hammer-beam truss, in which the dotted lines show the lines of resistance and BB being the level of the tops of the walls, it will be seen that the pressures are directed to the foot of the curved brace. The late Mr. G. E. Street, in his paper on English woodwork, read before the Royal Institute of British Archi

:cts in 1865, refers to the curved brace which distinguishes
1e roof of Westminster Hall from other examples (fig. 76)
s a source of strength to this type of roof.

A well-known hammer-beam roof for a small span is that
f Knapton (fig. 77), which is given, as also the scantlings of
1e various parts. During the fourteenth and fifteenth cen-
iries especially, many examples of this type of roof exist,
1d several are well executed in Brandon's "Mediæval
.oofs," to which the student is referred, as also to exist-
1g examples, as Westminster Hall (68 ft. span), Hampton

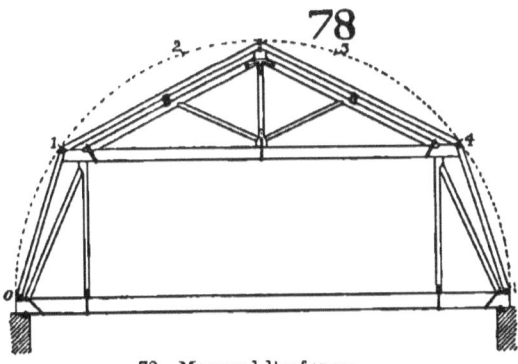

78. Mansard Roof-truss.

ourt (40 ft. span), Eltham Palace (36 ft. 3 in. span), and to
any examples in the country, Norfolk and Suffolk being
pecially rich in such specimens of carpentry.

Leaving the Middle Ages, we find that in the Renaissance
:riod in France *Mansard roofs* (fig. 78) were much used,
1d were so called after the name of the architect who in-
:nted them. It has always been condemned for its supposed
ant of beauty, but nevertheless has been largely used. Its
·igin may be derived from the desire to diminish the height
˙ the upper portion of steep roofs. The illustration (fig.
{) shows the method usually adopted in setting out this
of, as well as an ordinary type of construction, in which
e upper portion is merely a king-post truss. The next
ustration (fig. 79) shows the method for forming a dormer
ndow in this form of roof.

Philibert Delorme, a French architect, was apparently the
st to realise the saving in expense by using timbers

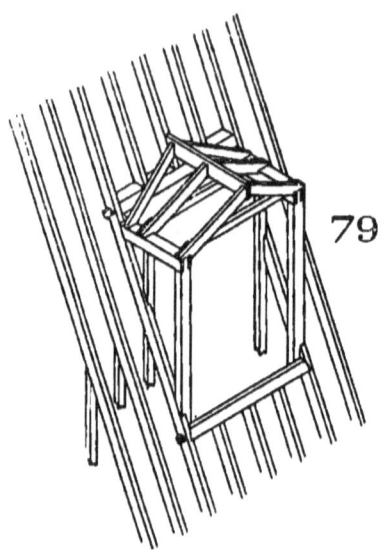

79. Dormer in Mansard Roof.

in small sections. His method was that of a semi-circular *plank truss* (fig. 80), which is made out of two thicknesses in pieces about 4 ft. long, laid together so that the joints of one set come in the middle of the length of the other. The planks for a span of 25 ft. may be 8 in. wide and 1 in. thick, and for 100 ft. 13 in. and 3 in. thick. These ribs are placed 2 ft. apart, and are connected longitudinally by ties, 1 in. to 1½ in. thick by 4 in. wide, which pass through mortises in the planks, to which they are pinned by keys, 1 in. by 1½ in., driven against the sides of the ribs. Even these may be dispensed with if boarding is nailed across the trusses. This system is useful for temporary exhibition buildings, and has, in some form or other, been largely used for the great exhibitions both at home and abroad. The truss may be stiffened by a framed brace at the abatements.

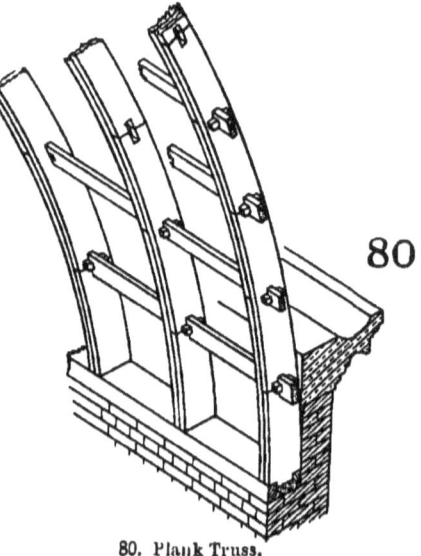

80. Plank Truss.

CHAPTER VII.

BRIDGES.

THE oldest Roman wooden bridge of which we have any record is that of Sublicius (from sublica, a pile), built about

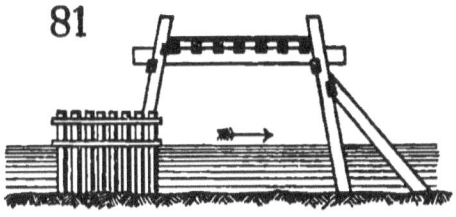

81. Cæsar's Bridge.

600 B.C., in the reign of Ancus Martius. This was a bridge, according to Polybius, that Horatius Coccles defended with so much valour against the army of Lars Porsena. Fig. 81 represents the supposed section of the celebrated bridge designed by Julius Cæsar to span the Rhine, and so quickly was it built that within ten days of the commencement of its erection his army had crossed it. The line struts supported the bridge against the force of the stream. Trajan's bridge over the Danube is represented in Fig. 82,

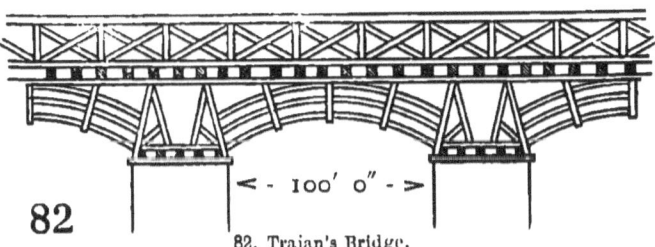

82. Trajan's Bridge.

which is copied from a bas-relief on the Triumphal Column erected to him in his Forum at Rome. Each arch was

100 ft. in the clear, and it will be noticed that its ingenious and pleasing design is eminently suited to such a structure.

Though at the present time bridges of large span are usually built of stone or steel, still in some cases, owing to financial or other reasons, the carpenter is called upon to erect an artificial elevated way in timber between two points separated by an intervening depression. In America, for example, there are few railways that have not, at some point on their permanent way, trestle-bridges formed of timber (fig. 83) spanning ravines; though from the oft-re-repeated news of their collapse, they seem perhaps somewhat out of place in railway construction. These calamities could, no doubt, be prevented by periodical inspections, and by spending a little more care and forethought on their maintenance. A committee, appointed by the Association of Railway Superintendents of Bridges and Buildings of the United States, have just issued their report, which, however, throws little fresh light on the subject, but recommends that more attention be given to the compression bearing resistance of timber across the grain, owing to the liability of heavy stresses to indent the timbers and thereby destroy the fibres, increasing the tendency to speedy decay.

The first point to be considered in the construction of a

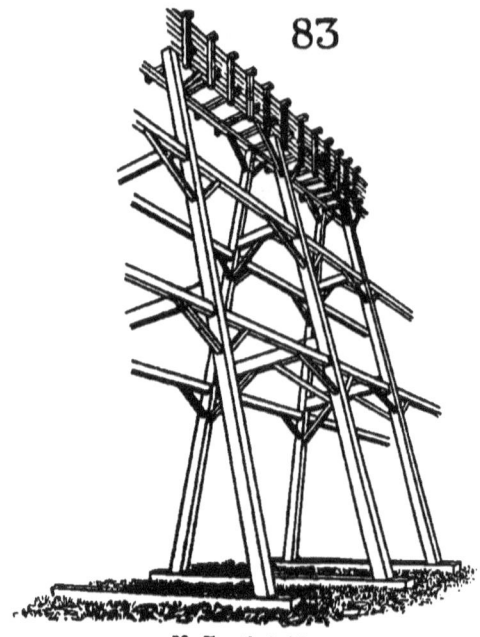

83. Trestle-bridge.

bridge is that sound and well-seasoned timber only be used, and that it is in sufficiently large scantlings, which must be well jointed together. In the event of spanning water care must be taken to calculate for the rapidity of the flow of the river, not only in its normal state but as likely to be increased by floods. An important consideration in designing bridges is to obtain the best situation having regard to local requirements and traffic, care being taken that there is sufficient means of access; further, it is always desirable to cross the stream at right angles if it is possible to do so. In crossing water a single span should, if possible, be used, because in the history of wooden bridges a far larger number are destroyed by the water beneath and by ice-flows, &c., than by weight exerted on its traffic surface.

Single spans can generally be constructed up to about 350 ft. or even 400 ft. No bridge is more satisfactory than a curved rib bridge in one span, and none can be stronger, for if well made it is more firm than a single piece of large dimensions, provided that wood is well selected and the direction of the grain is so arranged in the framing that the parts strengthen and support one another. The following table may prove useful to the student.

Table of the least rise for different Spans (Tredgold).

Span in Feet.	Least rise in Feet.	Span in Feet.	Least rise in Feet.
30	0·5	180	11
40	0·8	200	12
50	1·4	220	14
60	2	240	17
70	2½	260	20
80	3	280	24
90	4	300	28
100	5	320	32
120	7	350	39
140	8	380	47
160	10	400	53

The settlement of wooden bridges is generally taken as about 1 in 72.

Piers, for supporting bridges in cases where the river is not too deep or the current too strong, are made by driving in a single row of piles parallel to the current of the stream. These piles should be about 12 in. square and some 3 ft. apart. Fig. 100 is an illustration of a pier on this principle. When, however, greater strength is needed, there should be a double row of piles well braced together.

In constructing a bridge over the Severn, near Shrewsbury, Telford adopted a very successful method. He constructed piers consisting of an upright timber frame, with horizontal grated-framing attached, which formed its base. The framing was then sunk to its proper position on a level bed and short piles were driven through the framed-grating to keep it secure, the lower portion being filled with gravel and small stones, kept in their place by planking. Piles should have their lower end brought to a point and shod with iron, and should be provided with a hoop at the upper end to prevent any tendency to split from the blows of the *monkey*. In cases where more than one row of piles is used, it is usual to protect them by *sterlings*, which are simplest when composed of two rows of piles, converging to a point or cut-water. These piles are driven so that the point (on plan) is placed, like a wedge, against the stream, and the whole of the face is generally protected by iron to resist the force of ice-floes or other floating material.

TYPES OF WOODEN BRIDGES.

We may roughly divide bridges into five classes :—

I. *Flat lintel and cambered lintel.*
II. *The braced lintel.*
III. *Trussed girder bridges.*
IV. *Suspension bridges.*
V. *Curved-rib bridges.*

I.—*The flat lintel* is obviously the simplest form of timber bridge and may be strengthened by making it deeper in the middle than at the ends. The best form of *cambered lintel* is that shown in fig. 88, which may be composed in two pieces, or have several bolted together.

In the former case, the lower may be of a rectangular shape, the upper one tapering from the centre to both abutments.

II.—*The braced lintel* is the next modification from the flat lintel, and generally consists of the introduction of struts or braces below the level of the roadway. In the

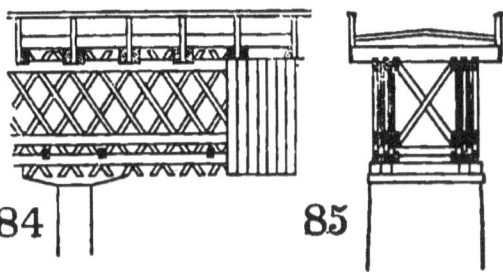

84. Lattice Girder Bridge.   85. Ditto.—Cross Section.

Middle Ages, when wooden bridges had to be formed over the principal rivers to connect the highways, they were generally constructed on this principle, with piers about 20 ft. apart, consisting of one or more rows of piles, protected by a kind of *jetta*. In rivers which were tranquil and not liable to high and rapid floods the augmentation of the piers did not so much signify, provided it did not unduly contract the flow of the river or interrupt the navigation, and

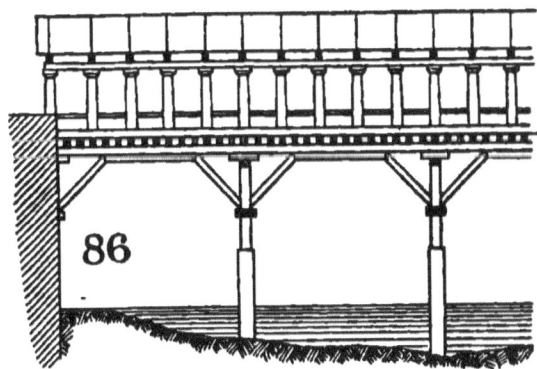

86. Bridge over the Brenta, by Palladio.

at the present day there are many cases where similar but lighter and wooden bridges may be used with much advantage, especially where the piers can be kept of slight dimensions. The Mediæval bridge-builder, however, often paid no regard to floods or currents, and consequently his numerous piers would collect the floating *débris* and become unable to withstand the pressure of water in time of high floods.

The bridge over the Brenta by Palladio (fig. 86) is a good example of the application of this principle, and, where practicable to erect a bridge of this kind, it would be difficult to find a more simple and effective method.

III.—*The trussed girder bridges* were much in favour with Palladio, who was amongst the first of modern

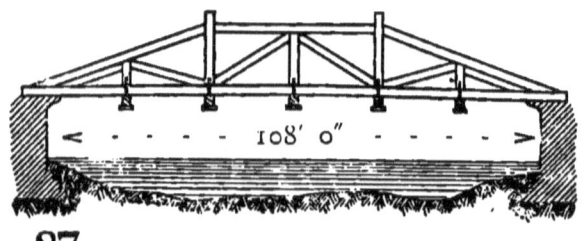

87. Bridge over the Cismone, near Bassano.

88. Flat Lintel Bridge.

architects to attempt to design bridges which did not require numerous piers in their construction. The celebrated bridge over the Cismone, near Bassano (fig. 87), was one of the best efforts of this architect, assisted by an able carpenter named Martino; its span is 108 ft. The transverse girders below, which are strapped up to the trusses and form the roadway, are 12 in. by 12 in., and the trusses are formed out of 12 in. by 9 in., and themselves form the parapet. It will be noticed that the construction assumes the form of the familiar queen-post roof-truss, a form which is eminently suited to bridge-building. Fig. 89 represents the centre bay of the foot-bridge designed by Peter Nicholson to cross the Clyde at Glasgow in the year 1858. It was specially designed with the view of admitting a certain class of vessels to pass beneath it. A clear span of 42 ft. was considered sufficient, making nine bays in all. It will be seen that a queen-post truss, formed as a railing, spanned from pier to pier. The breadth of the footway is about 8 ft., and the land abutments are strong masses of masonry. It will be noticed that each point has a lintel between it and the tie-beam of the truss, which tapers off

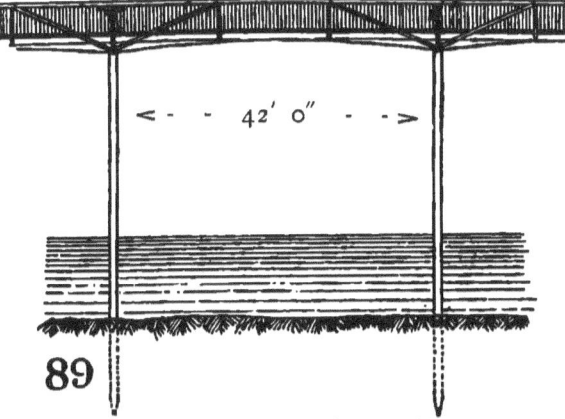

89. Foot-bridge across the Clyde at Glasgow.

towards the centre of the bay, thus assisting the stability of the bridge on the cantilever principle. The form of the footway is a flat parabolic curve. This bridge met with the approval of Rennie and Telford on account of its simplicity, strength, and appropriateness for its intended purpose. The railway bridge at Richmond in the United States, which is 2,900 ft. long, and is formed in spans of 150 ft., consists simply of two latticed girders 10 ft. deep, braced together with short transverse beams, on the upper surface of which are placed the railway lines. Figures 84, 85 represent a similar lattice girder bridge invented by Ithiel Town. It is suitable for spans up to 150 ft., and is much used on American railways, one great advantage being that only small scantlings are required in its construction. The lattice is formed of fir planks about 12 in. by 3 in., connected at their intersection by 1½ in. oak pegs. The depth of the lattice is generally about one-ninth of its span. The two ribs under each side of the permanent way are stiffened by cross timbers about 12 ft. apart, diagonal braces being inserted below. Stone abutments for these bridges find favour with the American engineer. The economy of this system is evidenced by the fact that the railway bridge at Richmond, United States, cost only between £8 to £9 per lineal foot for its construction, and it is 60 ft. above the water.

IV.—*Timber suspension bridges* are not in general use

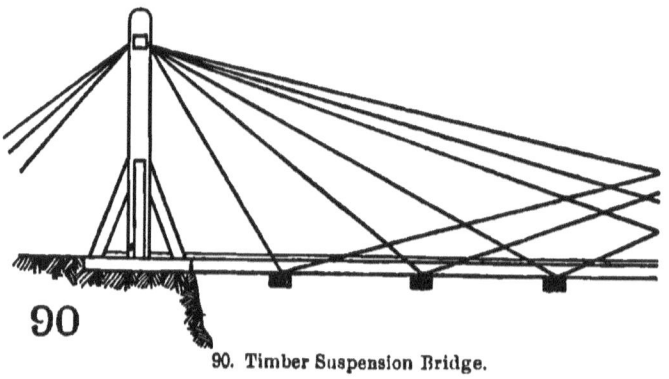

90. Timber Suspension Bridge.

and can be recommended only in very exceptional cases. Fig. 90 shows a simple yet ingenious method, which has been used with some success.

V.—*Curved-rib bridges* are suitable for large spans, and especially in places where, owing to the rapidity of the current, or liability of danger to piers, through floating masses or sudden and strong floods, one span is found desirable.

The radius of curvature of any bridge without horizontal ties should be equal to one-fifth of the height of a column, 1 in. square, of the material of which the bridge is composed, whose weight would crush 1 square inch of such material. For example, it has been found by experiment that a square inch of oak crushes at 5,147 lbs. One-fifth of this weight of oak in a column 1 in. square would be 2,950 ft. high. Consequently, the radius of curvature in an oak bridge should not exceed 2,950 ft., and, in the case of fir, 3,000 ft. should be the limit.

In a bridge with horizontal ties of timber, the bridge being, generally speaking, at least double the weight of the foregoing, the radius of curvature must necessarily be smaller, and is usually taken at half the above radii, viz., 1,475 for oak, and 1,500 ft. for fir. This, however, is without taking the weight of the roadway into consideration, which adds to the weight of the structure, but does not contribute to its support. This fact will vary the radius already determined, and may be found by the following formula:—

W = whole weight of bridge.
w = weight of support (frame only).
r = radius of curvature already determined.
θ = The final constructive radius required.

$$\theta = \frac{w \times r}{W}.$$

To find the sectional area at the crown for each rib of a curved rib bridge, let

β = sectional area in inches.
S = span in feet.
W = gross distributed load in lbs.
n = number of ribs.
R = rise in feet.

$$\text{Then } \beta = \frac{S \times W}{8,000 \, R \times n}.$$

For example, to find the sectional area of each rib at the crown of a bridge 100 ft. span, and 10 ft. rise, capable of supporting a load of 300,000 lbs, the curved ribs being three in number.

$$\beta = \frac{100 \times 300,000}{8,000 \times 10 \times 3}$$

$$\beta = 125 \text{ sq. in.}$$

Such a rib can be made up approximately of four pieces of stuff, 11 in. by 3 in., or two pieces, 11 in. by 6 in.

The sectional area of the ribs of the abutments will be slightly in excess of that at the crown owing to the greater pressure. Timber arches, unless made perfectly rigid by proper bracing, are not adapted to a variable load, such as a train in motion.

A curved rib bridge was designed by Palladio, and shown in fig. 93. It was probably the first idea of constructing a bridge on the principle of what may be described as *framed voussoirs*, similar to arch stones, and it is a principle which has been adopted in some instances both for timber and iron construction. It is contended by some authorities that,

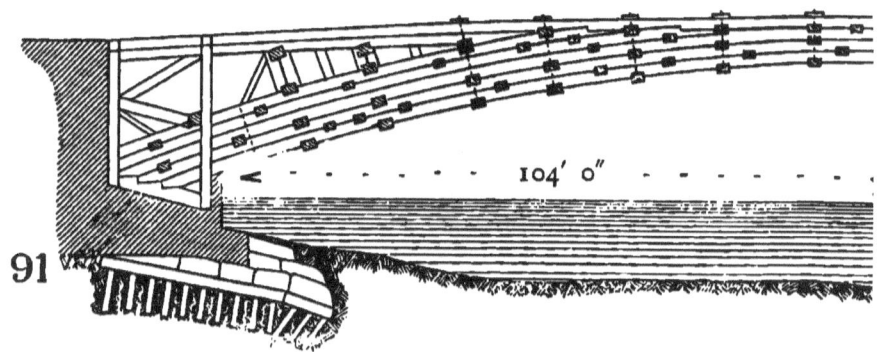

91. Bridge of Bamberg on the Regnitz.

owing to the shrinkage of the timbers and the vibrations caused by varying loads, the framing is continually endangered. The bridge of Bamberg on the Regnitz (fig. 91) in Germany, which was designed by Wiebeking in 1809, is a very good example of a close-ribbed bridge, erected to replace a former stone structure, which, by its piers, had so contracted the waterway that it was overturned by a flood. Consequently, the timber structure was designed to cross the river in one span of 208 ft., with a rise of 17 ft., and the width of the roadway was fixed at 32 ft. On looking at the cross section (fig. 92) which is to a larger scale, it will be seen that in the centre of the bridge three composite ribs are placed side by side, the centre one being five beams in depth at the abutments, which, however, are reduced to three at the centre of the span; the other two on each side of the centre consist of three beams throughout the span. On each side of the bridge two composite ribs are shown which consist of five beams at the abutments, reducing to three at the centre. The timbers, which were built into the abutments, were steeped in hot oil and covered with sheet lead. The cross ties and plates were constructed of oak, and the ribs and joists of fir. Price designed a simple but very useful bridge which could be carried to any desired place and erected at short notice; it can usually be used for spans up to about

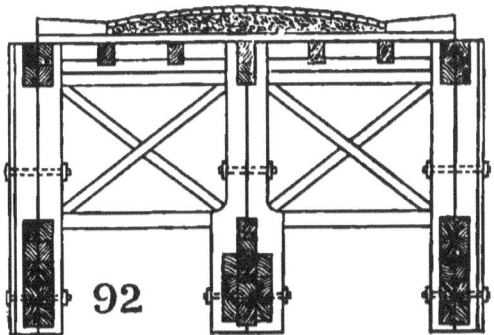

92. Bridge of Bamberg—Cross Section.

40 ft. It consists of two curved ribs, one being 12 in. by 3 in., and the other 9 in. by 3 in., placed side by side, and breaking joint; these were connected by two ribs of similar dimensions forming the other side of the bridge, the connecting joists forming the roadway. The span was divided into five parts so as to make it portable, and the whole structure was fitted together by mortising the connecting joists into the ribs and keying them up. The celebrated wooden bridge over the Portsmouth River in North America (fig. 94) is a scientific development of Price's idea. This bridge has a span of 250 ft., and a reference to this illustration will show how ingenious is its

F

construction, and how, by the system of wedging up its parts, the whole structure is rendered homogeneous. Figs. 95, 96, 97, 98, 99 show details of the wedges and keys.

Roadways.—The ordinary width of the roadway of a bridge varies from 18 ft. to 45 ft. for vehicular traffic, and from 5 ft. to 10 ft. for foot traffic. A width of 18 ft. just

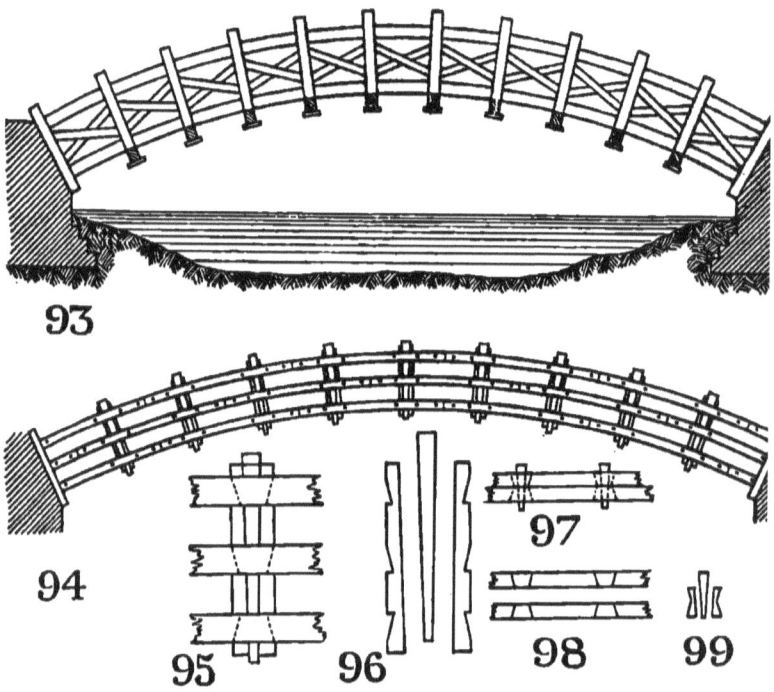

93. Curved Rib Bridge, by Palladio.
94. Wooden Bridge over Portsmouth River, N. America.
95 and 96. Detail of Wedges.
97, 98, and 99. Detail of Keys.

admits of two carriages passing one another, so that the carriage-way should vary by a multiple of nine. It is better not to lay the planking directly on to the principal beams, but to lay cross-joints, which would admit of a freer circulation of air. The road itself may consist of gravel and tempered clay, the latter to bind the gravel together, of about 12 in. in depth at the centre and 9 in. at the side. Separate footpaths must be made for passengers from 2 ft. to 6 ft. wide, and means for carrying off the water must always be provided. Occasionally—as, for instance, on the bridge of boats at St. Petersburg, which in winter is swung back against the bank to allow sleighs to pass on the ice, and on the bridges crossing the Golden Horn at Constantinople—it is found expedient to make a roadway of planking. This should consist of a double layer, the top one placed across the bridge to enable the horses to obtain a foothold; it is often treated

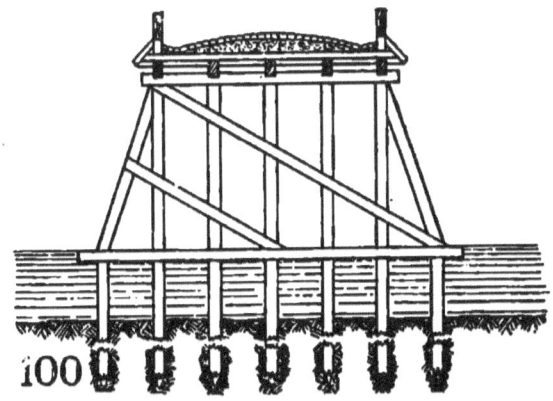

100. Piles forming a Pier.

with a coating of tar and sand. Bellidor recommended that bridges be paved, which he contended added greatly to their durability. Parapets should not be less than 3 ft. 6 in. high.

Scantlings of bridges should be calculated to allow for a crowded procession to pass over safely—the greatest strain it will be subjected to. Such a load is about 120 lbs. per super, which, if added to the weight of the framing and the roadway generally, makes a total of about 300 lbs. per foot.

CHAPTER VIII.

SHORING AND STRUTTING.

THIS important branch of the craft demands great skill and judgment, occasionally leaving but little time for calculation, and sometimes involving no inconsiderable amount of personal risk. The subject has been much neglected in its theoretical aspect by the architect, who too often leaves the scheme and system of shoring almost entirely to the discretion of the builder's foreman. This is the more astonishing as, in one case at least, a verdict of manslaughter was recorded against an architect, after the fall of a house just outside the western boundary of the City of London. Though much must be left to the almost incessant and intelligent watching of the builder and his workmen more especially in needling operations, still, the architect should give personal supervision both to the design and erection of all temporary wooden supports.

SHORING may conveniently be treated under the following heads :—

I. *Raking Shores.*
II. *Flying, Horizontal, or Dog Shores.*
III. *Needle Shores and Underpinning.*
IV. *Special Forms of Shoring.*

I. *Raking Shores* are used when it is required to support a structure either because it is too far away from any other building to use a flying shore, or on account of the owner of the adjoining building objecting to his building being used for the purpose of supporting that of his neighbour. The simplest form of raking shore is that shown in fig. 101, which may be used to support a one-story building. A is the *raking shore*, W is the *wall-piece*, 9 in. by 3 in., and is sufficiently long to take the foot of the strut, S. A rectangular hole, 6 in. by 4 in. or 6 in. by 6 in., is cut in the wall-piece at A, into which a *needle*, N, which fits the hole is driven. This is about 12 in. long, and is placed in the wall from which a header has been previously removed

Where possible this should be placed immediately under the floor level. A *cleat*, C, is nailed to the wall-piece above the needle to assist the latter in resisting the upward thrust of the shore. SP is the *sole-piece* let into the ground, upon which the shore abuts, and to which it transmits the lateral pressure of the wall. In the case of soft or untrustworthy ground, the sole-piece should rest on planks, as shown at

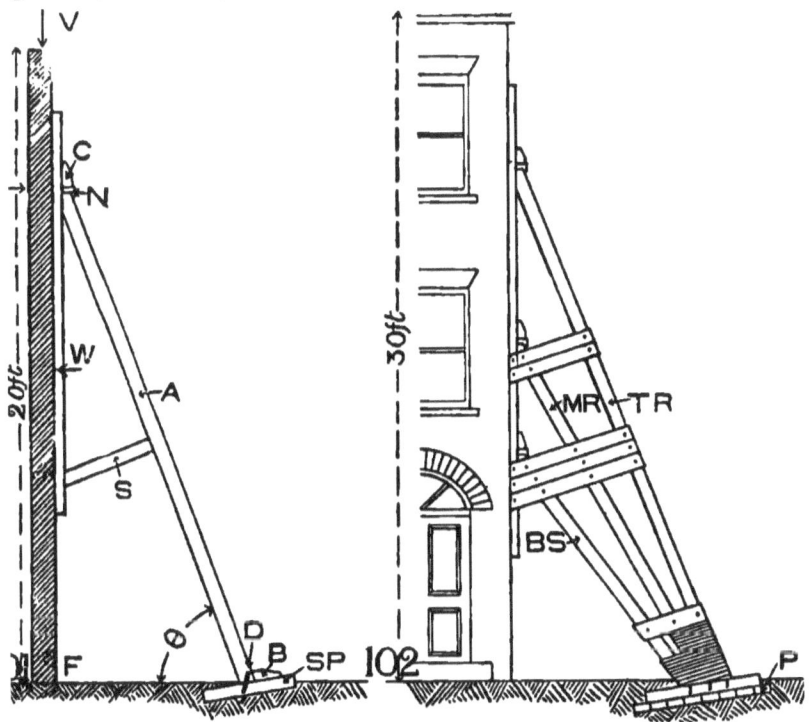

101. Raking Shore: simplest form.   102. Raking Shore: most frequently used.

P in fig. 102, so as to distribute the weight. The shore is kept in position on the sole-piece by an iron dog, D, and by a cleat, B.

Fig. 102 represents the raking shore most frequently used in London, and is a natural development of fig. 101 applied to a three-story building. The longest (outside) shore, TR, is called the *top raker*; the centre shore, MR, is called the *middle raker*, and the lowest, BS, *bottom shore*. The wall-piece is of course considerably longer. Three rectangular holes are made in it, and the needles inserted and cleats used similar to those previously described. Instead of the struts shown in fig. 101, boards 1 in. thick and 6 in. to 9 in. broad are nailed across the shores and on to the wall-pieces

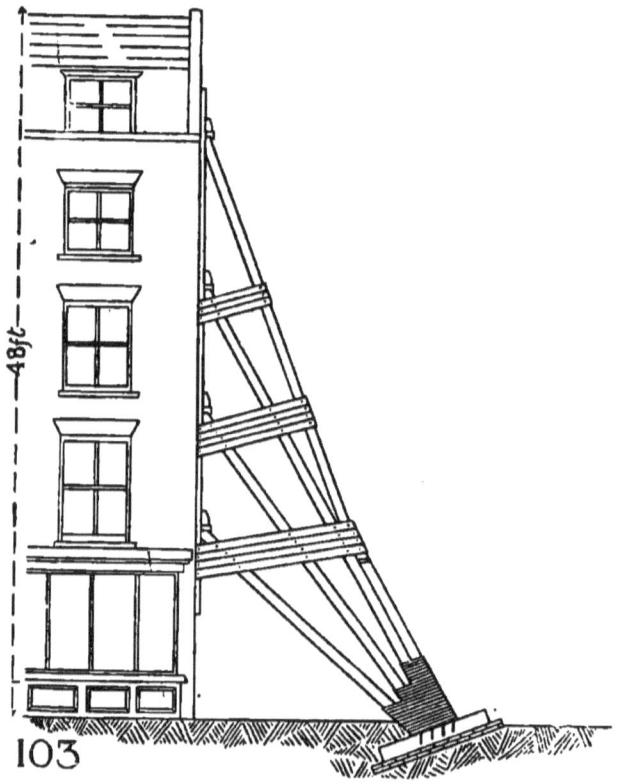

103. Raking Shore: adapted for very high buildings.

at places adjacent to the needles. The shores at their base are generally either treated in a similar manner or else bound round with hoop-iron.

Fig. 103 shows a system of shoring adapted for a four-story building, with an additional attic story. In this case, which is similar in almost every other instance to the preceding example, the *top raker*, or *riding shore*, as it is more often called in this form, is constructed in two pieces, owing to the inability to procure timber of sufficient length.

Where, however, timber of the requisite length can be obtained, it is almost always desirable to form the shores in one piece, as the power of timber to resist compression is materially affected by any cross strain.

In the case of a frail structure, it is sometimes considered that the manipulation of large timber shores may impair the building, and in such cases it is best to have the rider in two pieces. When this is done, it is best for the top shore to be divided from the lower shore, which should be of the same scantling and resting on the sole piece, by a pair of oak folding wedges.

The design of raking shores may vary slightly in each individual case, but in shoring, perhaps more than in any other branch of carpentry, the same general principles have to be observed. The best angle of inclination for raking shores is generally considered to be 40 deg., but as this involves the use of so much room, they are rarely seen at a less angle than 60 deg. or 70 deg., though, of course, the lateral thrust varies in proportion as the angle is increased. It may be taken as a general rule that one raker is required for every story in an ordinary building, omitting to count in the attic story where one exists, and the head of each raker is generally placed against the wall-piece at the level of the floor-joists. Shores should not be placed at a greater distance than from 12 ft. to 15 ft. apart, and if placed nearer together the scantlings may be reduced. The following table will give a general idea of the scantling generally used for each shore when the angle of inclination is between 60 deg. to 75 deg.

	Ft.	Ft.	In.	In.		In.	In.
Walls—	15 to 20 high		4 by 4	or	5 by 5
,,	20 to 30 high		9 by 4½	or	6 by 6
,,	30 to 40 high		12 by 6	or	8 by 8
,,	40 to 50 high		9 by 9	or	8 by 8
,,	50 and upwards		12 by 9	or	11 by 9

The following formulæ may be found of use in calculating the compression the shore has to resist, and in arriving at the scantlings:—

(a) Let H = the horizontal thrusts in cwts. (see fig. 101).
W = the weight of the wall in cwts.
t = the thickness of the wall at the base in feet.
NF = the distance of the head of the shore from the ground.

Then $H = \dfrac{W \times t}{2(NF)}$

(b) Let V = the vertical force in cwts. (see fig. 101).
θ = the angle the shore makes with the horizon.
W = weight of the shore in cwts.

Then $V = H \tan. \theta - \dfrac{W}{2}$.

(c) Let C = the compression down the shore in cwts.
θ = the angle the shore makes with the horizon.
Then $C = V \sin. \theta + H \cos. \theta$.

(d) Let S = safe load of shore in cwts.
$a = 15\cdot 5$ for fir.
d = width in inches of shore.
e = length in feet.

Then $S = a \times \dfrac{d^4}{e^2}$.

NOTE.—This latter formula, of course, is for square timbers; where one side is greater than the other, take d as the lesser and multiply S by $\dfrac{\text{the greater side}}{\text{the lesser side}}$.

The above calculations are for the top raker. The compression down the lower shores is, of course, greater; but this is considered to be counterbalanced owing to the fact that they are shorter, and their power of resistance is, therefore, greater; it is also more convenient for construction to have them of a uniform size.

With regard to the angle at which the *sole-piece* should be placed with the raker, it is found that the resultant force of V and H (fig. 101) acts in a direction outside the angle made by the shore with the horizon, consequently the sole-piece must not be placed at right angles to the shore itself, but at right angles as near as may be to the direction the resultant may take, which, therefore, is always less than a right angle.

Fig. 104 is a detail showing the notching of the head of the raker, with the needle and cleat towards the upper end of the wall piece. In the event of a shore being required for any length of time, this notching should always be done; where, however, the shore is of a very temporary nature, it may be secured by iron *dogs*, which are also used for securing the feet of the shores to the sole-piece, for keeping the rider firmly on its lower shore, and also for keeping the various rakers in position. Fig. 105 shows the correct form of dog at A, and the incorrect one at B. In the former case the driving of the dog home brings the shores together, and in the latter case these are driven apart.

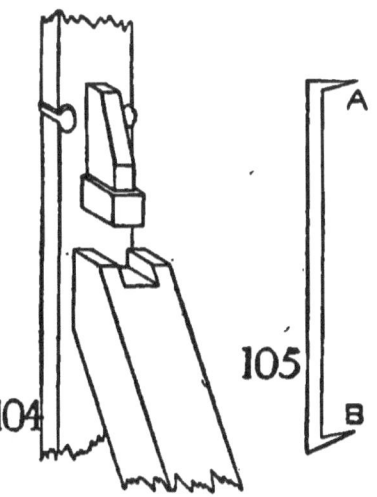

104. Detail showing notching of head to needle.
105. Correct and Incorrect Form of Dog.

A square hole is often cut on the base of raker into which a crowbar is inserted, by means of which it is gently levered into its place on the sole-piece. Fir is the best timber to use for shores, and oak should be employed for the wedges. All joints should fit well, and the timbers should have a good firm bearing on the sole-piece.

Fig. 106 illustrates a very strong combined shore which might be found very useful for some purposes, such as shoring up a retaining wall.

Referring to the triangulation of shores, Viollet-le-Duc says that "they should always form a triangle or portion of a triangle, for the reason that a triangle can never be thrown out of shape; when braced, shores which are not parallel present an entirely homogeneous resistance, whereas if they are parallel, they will become bent, however well braced they may be." Raking shores have been used for the purposes of pushing a wall back to its original vertical position, by placing powerful screwjacks under the sole-pieces. This of course could only be effected where the

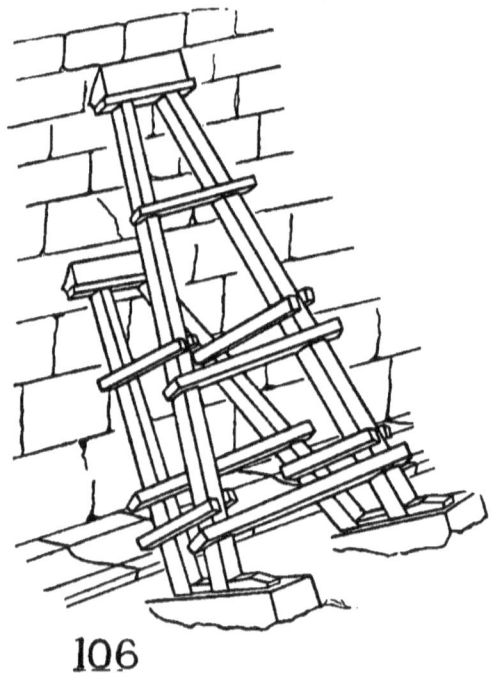

106. Strong Combined Shore.

foundations are sound, everything being disconnected and shored up on the other side; it is not a method to be recommended.

II. *Flying Shores*, also called *dog and horizontal shores*, are very convenient to use when the distance between the two adjoining houses does not exceed 33 ft., beyond which length Dantzic fir cannot easily be obtained.

The usual method of erecting these shores is shown in

fig. 107. Two rectangular holes are cut in the centre of each of the wall-pieces, and the needles, N, are then inserted, as previously described for raking shores, and a cleat, C, is nailed below them to help sustain the pressure.

Oak wedges, OW, are driven between the wall-piece and the horizontal strut, and above the needles. The raking struts, RS, help to distribute the pressure and to stiffen the main strut. They rest against straining-pieces, SP, nailed to the upper and under sides of the horizontal strut, and against cleats, C, on the wall-pieces. Sometimes needles are let in before the cleats are fixed, thus affording additional security, and oak wedges are used, OW, to tighten them up to the straining-pieces.

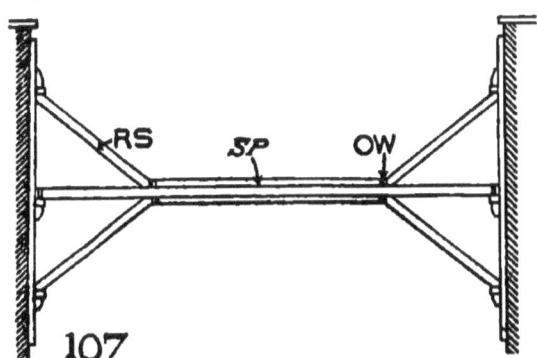

107. Flying Shore.

In flying shores, the thrust may be calculated as follows:—
Let H = the thrust in cwts.
  W = the weight of the wall in cwts.
  t = thickness of the wall at ground in feet.
  AF = the distance in feet from the ground to the point at which it is desired to find the thrust.

Then as before $H = \dfrac{W \times t}{2AF}$

Of course, the higher the shore is up the wall, the less will be the compression, and the lower the shore is placed, the stronger it must be. Very often more than one flying

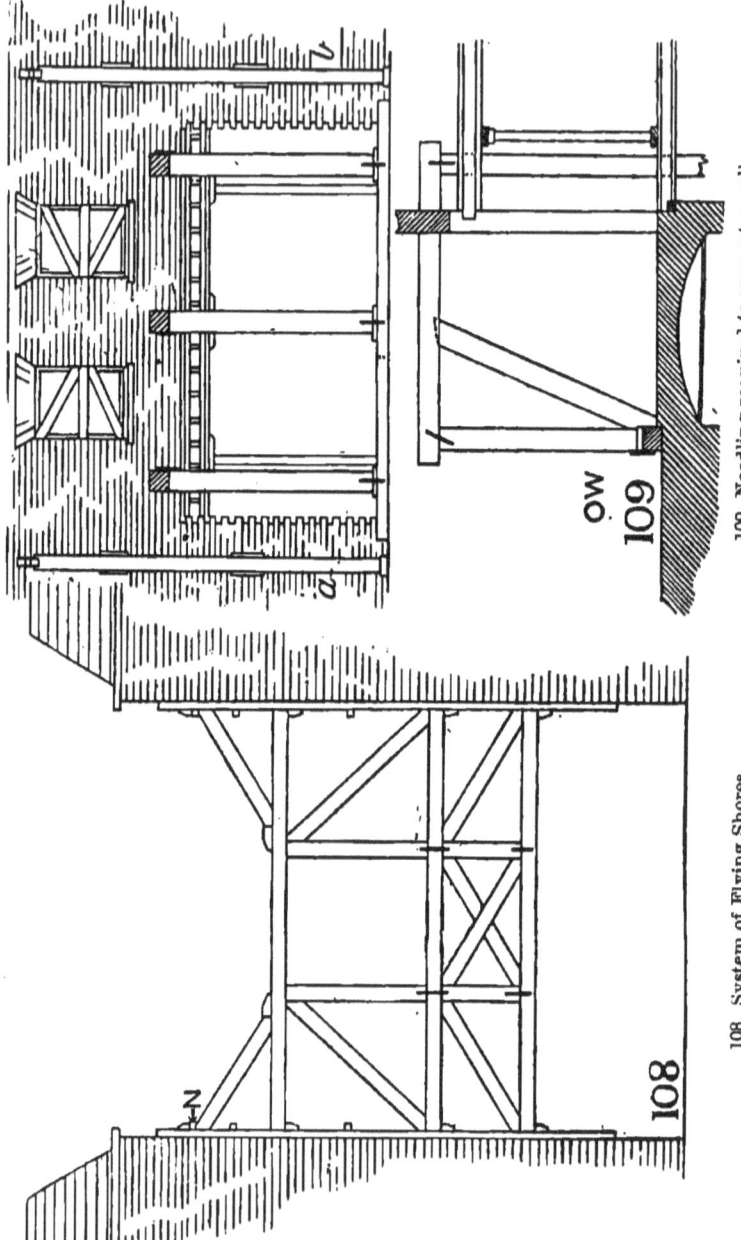

108. System of Flying Shores

109. Needling required to support a wall.

shore is required, and fig. 108 shows a system of flying shores, one above the other, with a continuous wall-piece. This latter, if the wall is of an uneven character, should be firred up behind, so that it may have an even bearing throughout. The following scantlings are usually considered to be adequate :—

For spans not exceeding 15 ft., raking struts, 4 in. by 4 in.; horizontal struts, 6 in. by 4 in.

For spans from 15 ft. to 33 ft., raking struts, 6 in. by 4 in. to 9 in. by $4\frac{1}{2}$ in. ; horizontal struts, 6 in. by 6 in. to 9 in. by 9 in.

The above scantlings apply to shores placed one-fourth of the total height from the top of the wall, and each bay of shoring should not be more than 12 ft. 6 in.

Flying-shores have the advantage over raking-shores of taking the direct thrust, and of not interfering so much in building operations.

III. *Needle Shoring and Underpinning* differ from the previous methods described in that the foundations are no longer relied upon, and the wall under treatment has to be held up in the air without any direct support from the ground. Rectangular holes are cut in the brickwork, through which balks of timber, generally about 12 in. by 12 in., are passed, and these are supported at each end by vertical posts of similar scantlings, resting on sole-pieces, which are usually laid on timber platforms on the ground. Oak wedges are then inserted under the uprights, and on being driven home, these force the needle tightly against the upper side of the brickwork in the cavity. These needles should not be placed more than about 6 ft. apart, and when all are in position they support the wall so that the lower part may be removed, and, as is often the case, a shop front or other structure may be inserted underneath.

Fig. 109 represents the needling required to support a wall during the insertion of a girder for some such purpose as indicated above. If the wall is perfectly sound, no raking shores would be required ; but if this should not be the case, it would be advisable to have a raking shore at *a* and *b*, and one between the windows ; in any case, the windows should be shored up as shown. The needles should be inserted above the floor level so as to have a grip on the solid brickwork, and, in the case of there being a basement, the vertical support should go straight through the ground floor to the solid ground beneath.

Care should be taken to support all floors, chimney breasts, piers and corbels in the walls to be needled, and these latter should be so arranged as not to interfere in any way with the insertion of bressumers, girders, or stanchions used in the alteration. Whole timbers should nearly always be squared, for though in most cases they are considerably stronger than necessary, still the slightest deflection might cause small fissures in the upper structure. The following formulæ are generally recognised as sufficient :—

To find a scantling of the needle that will sustain a given safe load in the centre :—

Let $L =$ length of bearing in feet.
$W =$ weight to be sustained in pounds.
$A = \cdot 01$ for fir and $\cdot 013$ for oak.
$B =$ breadth in inches.
$D =$ depth in inches.

To find depth where breadth is known or determined upon :—

$$D = \frac{\sqrt[3]{L^2 \times W \times A}}{B}$$

To find breadth where depth is known or determined upon :—

$$B = \frac{L^2 \times W \times A}{D^3}$$

To find the scantling of vertical posts capable of sustaining a given compression in the direction of their length :—

Let $L =$ length in feet.
$W =$ weight to be sustained in pounds.
$A = \cdot 0013$ for fir, $\cdot 0015$ for oak.
$B =$ breadth in inches.
$D =$ depth required in inches.

$$\text{Then } D = \frac{\sqrt[3]{L^2 \times W \times A}}{B}$$

If a wall be sound except for its foundations, it can be restored to a perfectly good condition by firstly erecting raking shores to assist in supporting the wall. The ground on each side of it is then dug out for not more than 5 ft.

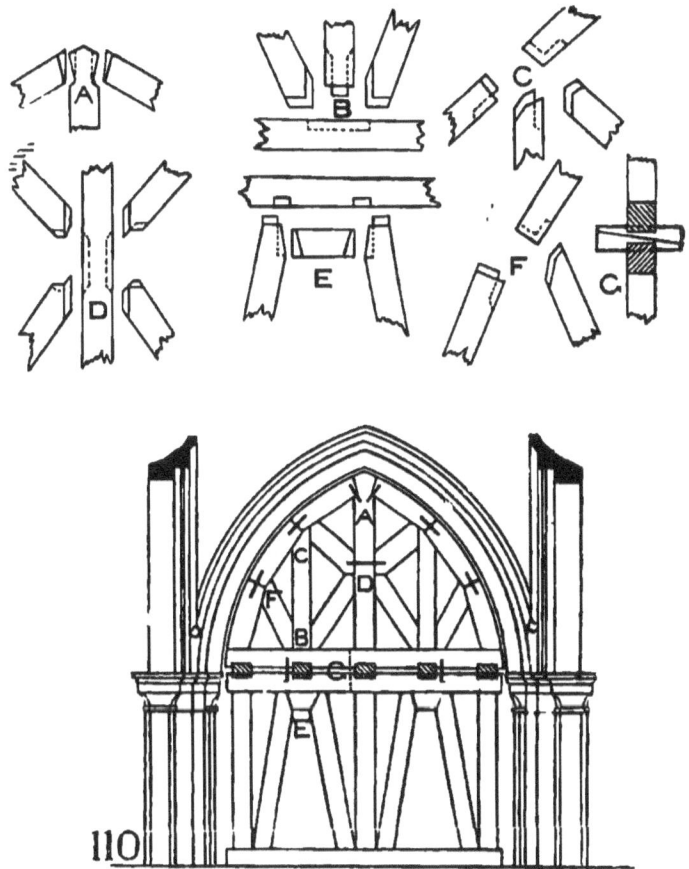

110. Special Form of Shoring used at Grosmont Church.

to 7 ft. at a time; the whole of the existing foundation is then removed and a new foundation built upon the solid ground. This is called *underpinning*.

IV. *Special Forms of Shoring* are often required which must be left to the skill of the architect and the builder to adapt or devise as the case may require, and it is not intended in this brief article to do more than refer to these special designs. Fig. 110 and those numbered A to G represent the system adopted by Mr. J. P. Seddon at the parish church of Grosmont in Monmouthshire. Fig. 111

111. Shoring used to support Cap of a Cylindrical Column.

represents the shoring suggested for the support of the cap of a cylindrical column which carries vaulting ribs in all directions. This would permit of the column being taken away and another built in its place. As to the timbering used in shoring, an allowance of one-third to one-half is usually made for waste or for reconversion to use.

CHAPTER IX.

CENTRES.

CENTRES are temporary structures, generally of wood, used to support arches of brick or stone, till they have settled in position and become consolidated. The qualities of a good centre consist in its forming a rigid support for the weight of the arch stones, without varying its form to any appreciable extent throughout the whole progress of the operation, from the springing of the arch to the laying of the keystone. Centres required for the erection of small and narrow arches may consist simply of a piece of stuff cut to the requisite curve of the soffit of the arch it is proposed to build, and kept in position by wooden supports or props, as shown in fig. 121. Where the span is over 3 ft. or

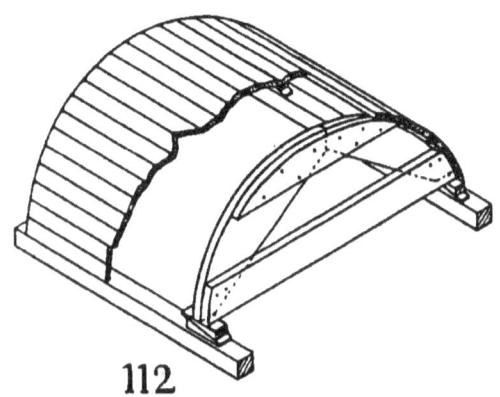

112. Centre for Small Span.

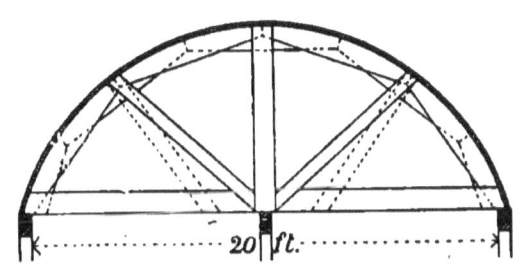

113. Centre for 20-ft. Span.

G

4 ft., centres are generally constructed as shown in fig. 112. A centre designed for a 20 ft. arch of three bricks in thickness is shown at fig. 113. It will be observed that it is designed on the principle of the king-post truss, though some architects prefer the struts to be placed at right angles to the arch, as shown by dotted lines. This centre is supported in the middle and at each end by 6 in. by 6 in. fir posts, and interposed between such posts and the horizontal tie are driven oak wedges so that when it is required to ease the centre, the wedges are tapped gently out of position.

In the case of tunnelling or other work involving greater depth, the ribs should be placed from 2 ft. to 6 ft. apart, the distance diminishing in proportion as the weight of the structure increases.

Before entering upon the question of the design and construction of centres for wide spans, it will be well to ascertain first what proportion the total weight of the structure the centre will have to sustain.

It has been found by experiment that a stone placed upon an inclined plane does not begin to slide until that plane has an inclination of 30 deg. from the horizontal line, and until such a stone would slide upon its base, it is obvious that the centre would not contribute to its support. Moreover, this *angle of repose* reaches as much as 45 deg., with a soft stone bedded in mortar, but 32 deg. or 34 deg. generally should be calculated for. As the courses approach more nearly to the keystone, so, of course, their weight on the centre increases until at last the centre has to support their whole weight. It is a useful fact to remember that when the plane of any joint becomes so much inclined that a vertical line, passing through the centre of gravity of the arch stone, does not fall within the lower bed of that stone, the whole weight of the stone may be considered as resting on the centre.

No absolute rule can be given as to the proportion of the whole weight which the centre will have to carry, as of course the ratio increases as the arch becomes flatter, but it may generally be safely assumed that the centre will not have to sustain more than two-thirds of the total weight of the arch. The following table is given by Tredgold where P represents the pressure and W the weight of any arch stone :—

When the angle which the joint makes with the horizon is .. } 34 degs. P = ·04 W.
,, ,, ,, 36 ,, P = ·08 W.
,, ,, ,, 38 ,, P = ·12 W.
,, ,, ,, 40 ,, P = ·17 W.
,, ,, ,, 42 ,, P = ·21 W.
,, ,, ,, 44 ,, P = ·25 W.
,, ,, ,, 46 ,, P = ·29 W.
,, ,, ,, 48 ,, P = ·33 W.
,, ,, ,, 50 ,, P = ·37 W.
,, ,, ,, 52 ,, P = ·40 W.
,, ,, ,, 54 ,, P = ·44 W.
,, ,, ,, 56 ,, P = ·48 W.
,, ,, ,, 58 ,, P = ·52 W.
,, ,, ,, 60 ,, P = ·54 W.

From the above it will be easy to estimate the weight upon a centre at any period of its construction, as well as the whole weight it has ultimately to sustain.

From the above remarks it will be obvious that it would be absurd to make the centre equally strong in all its points. Failures of centres have frequently occurred through the weight at the haunches having forced up the framing at the crown, and one of the chief objects to be borne in mind in designing centres is, that it should not undergo a change of form during any portion of the construction of the arch.

The designing of centres may, for convenience, be divided into :—

I.—*Centres supported from below at intervals.*
II.—*Centres in one span.*

I.—*Centres supported at intervals* should always be used where it is possible, as they are far more economical both in labour and material, and can more easily be depended upon to support the arch with rigidity. Fig. 114 shows the centre used for the Coldstream Bridge as designed by Smeaton, whose principal object was to divide the support equally under the weight, and at the same time to preserve such a geometrical connexion throughout the whole that, if any one pile or row of piles should settle, the incumbent weight would be supported by the remainder. He was also particular not so much to have his scantlings, which were of "East County" fir, as light as possible, but also to cut them with the least waste. This stone bridge was 25 ft. in width, and the span of the centre arch was 60 ft. 8 in. Fig. 115 is an illustration of the centring for the stone bridge built by Telford at Gloucester, in which the span

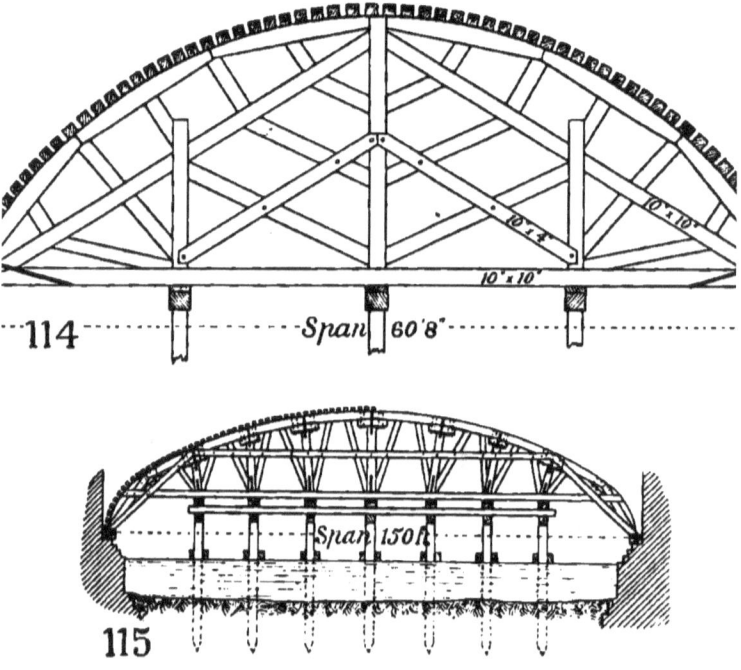

114. Centre used for Coldstream Bridge.
115. Centre used for Stone Bridge at Gloucester.

was 150 ft., and the rise 35 ft. The method of construction was as follows :—A level platform was prepared, on which the centre was struck out to the full size. The timber used was of Dantzic fir, in scantlings of about 15 in. square, and the piles carrying the centre were of Memel, shod with iron. Across the top of the piles a beam was laid, upon which the wedges were fixed. Each rib of the centre was then placed upon a scaffold made upon the top of the wedge pieces, and was raised into position by two cranes on the banks, aided by two barges on the river.

The scaffold was continued 30 ft. beyond the striking ends of the wedges, which were found convenient both for hoisting the ribs and for striking the centre. When the ribs were properly braced they were covered with 4 in. sheet piling, which had been previously used in the formation of the coffer dams. This centre was so successful that when

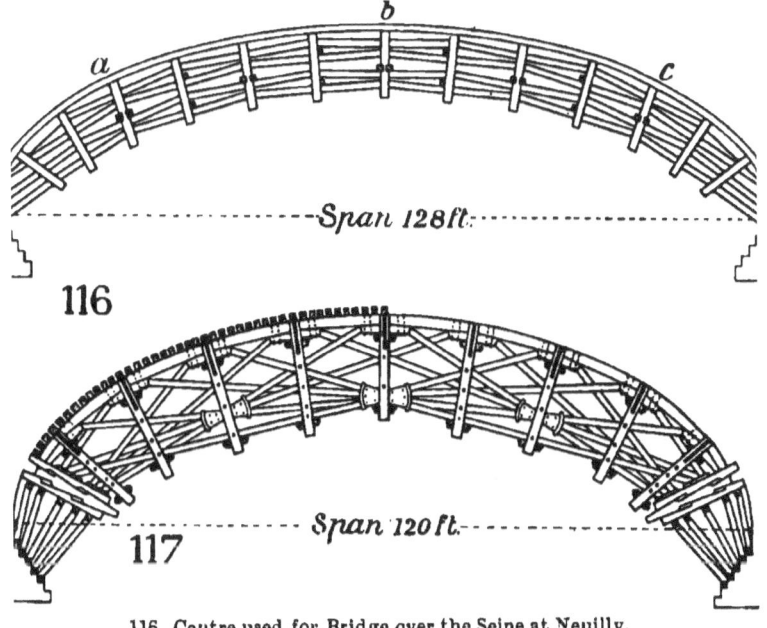

116. Centre used for Bridge over the Seine at Neuilly.
117. Centre used for Waterloo Bridge.

the arch was keyed, its sinking did not exceed 1 in., and it struck within the short space of three hours.

II. *Centres in one span* have to be constructed when intermediate supports cannot be used, owing either to the uncertainty of the current or to the necessity of keeping the water-way clear. Their execution is manifestly much more difficult, owing to the precautions which must be taken to counteract the tendency of the crown to rise when the load is placed on the haunches. This is forcibly illustrated in fig. 116, which is the centre designed by Perronet for the bridge at Neuilly. A slight examination will show that when it is loaded at a and c it must rise at b, and the strains produced by the weight resting on any point must have been very considerable, owing to the timbers being so nearly parallel, and the strains on the joints must have been excessive.

The centre has such a light appearance, and obstructs the stream so little, that it recommended itself to many, but it is contrary to the true principles on which centres should be designed, though it is well enough adapted to sustain an equilibrated load, distributed over the whole length; but it is certainly not calculated to support a variable load without alteration from its original form. Several other designs of a similar nature have been executed for other bridges, and have been found to be equally defective.

Fig. 117 represents the centring designed by Rennie for Waterloo Bridge, and it will be seen that, owing to the numerous cross-ties, a load placed in any position could not cause the centre to be raised without reducing the length of some of these cross-ties. It is contended by some that it is complicated, and that there is an excess of strength; but there is no doubt that it is well adapted for the purpose for which it was intended. It is a modification of the centre used for Blackfriars Bridge in 1760 by Milne, though improved in construction and form.

Fig. 118 is a form of centre recommended by Tredgold, consisting of three main trusses abutting against each other, from which it is apparent that, as the arch is built up from each abutment, the load on each of the haunches being equal, provided the central truss is sufficiently stiff to resist the pressure in the direction of its length, there will be no tendency to rise in the middle. It is easy to increase the strength as required. A centre on these principles may be executed for any span to which a stone bridge may be erected. If necessary, of course, the timbers may be lengthened as described in the article on joints. All very obtuse angles should be avoided, and to secure the utmost strength timbers should be subjected as little as possible to

CARPENTRY AND JOINERY. 95

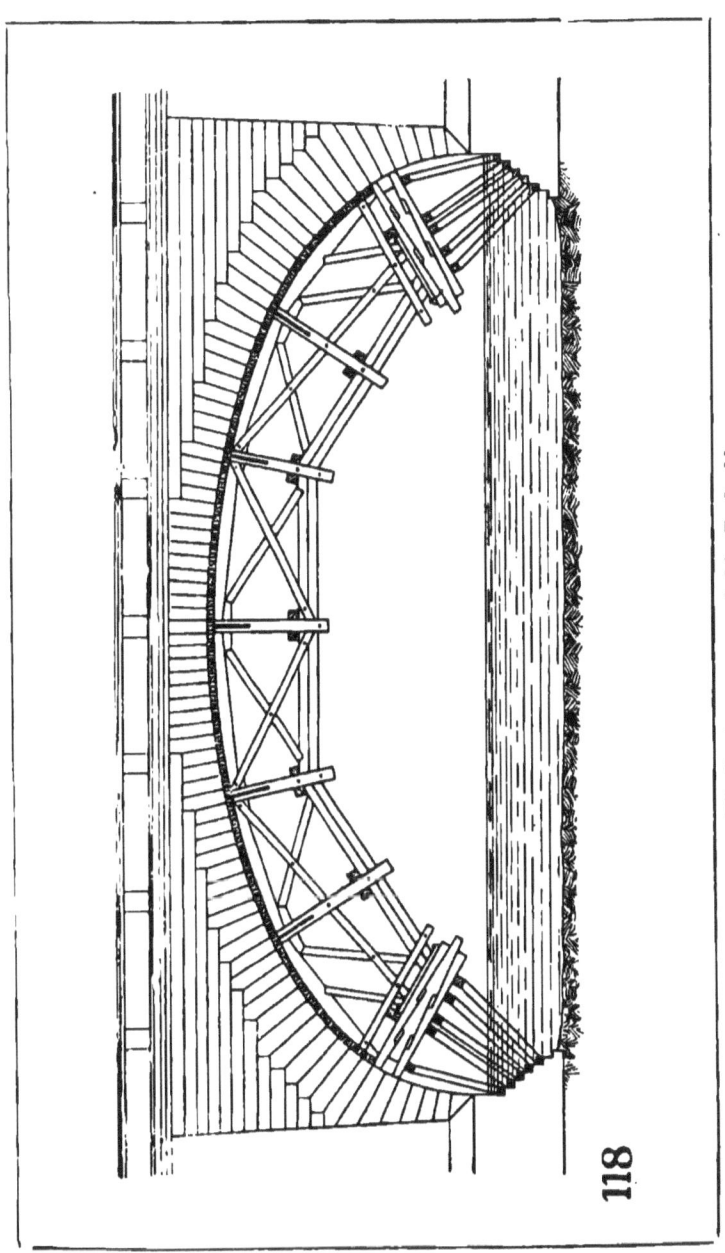

118. Centre recommended by Tredgold.

cross-strain. The girder principle, as explained in the chapter on "Bridges," may be advantageously used in many cases in the construction of centres.

Fig. 119 shows a very good centre, which is frequently used in the construction of tunnels. It will be seen that the framing is very similar to that of a queen-post truss, previously described in the article on "Roofs." The upper portions of the rib are made of 3-in. plank in two thick-

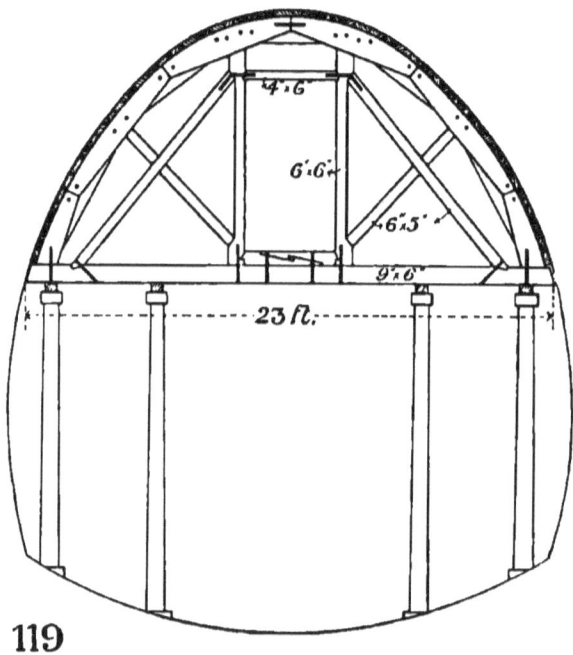

119. Centre usually used for Tunnels.

nesses, bolted together, and the ribs are usually about 3 ft. apart.

The two ribs nearest the excavation are constructed without tie-beams to avoid interfering with the raking struts.

The *construction of centres* will probably not require much explanation to those who have followed the chapter on "Joints used in Carpentry" and "Roofs," but the few following particulars may be of use:—The pressure upon any beam in pounds, divided by 1,000, gives the area of the timber in pieces, or that of the least abutting joint. Where it is possible, principal beams should abut end to end; when

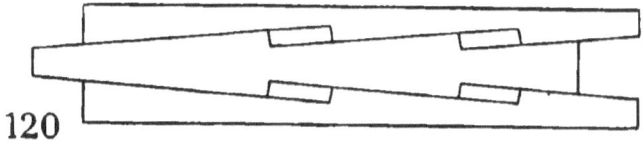

120. Detail of Continuous Wedge.

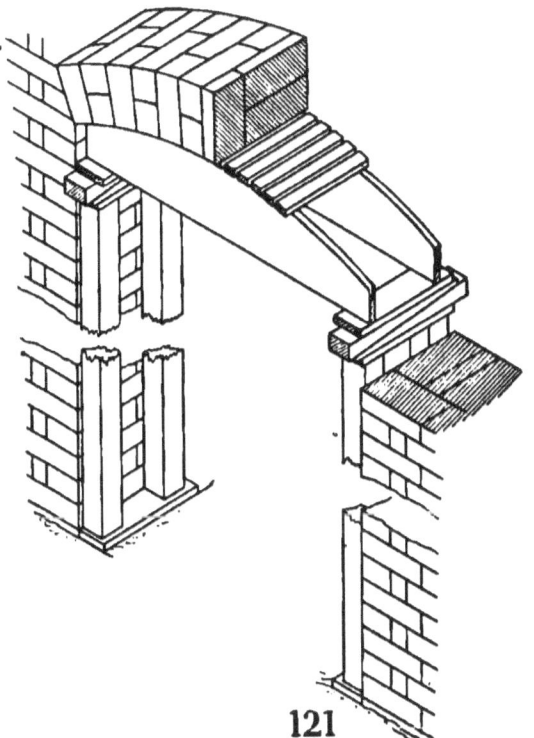

121. Centre used for Door and Window Arches.

they meet at an angle it is a good plan to let them into a cast-iron socket (see fig. 117). Timbers should intersect each other as little as possible, halving should be avoided, and those timbers which act as struts and braces should be notched upon the framing and be in pairs, one on each side of the frame, and well bolted together (see braces in fig. 119).

Ties should be used across the framing, particularly where many timbers meet, and diagonal braces between the ribs should be used to prevent lateral motion. The necessity for a sound and a rigid centre must again be urged, and as an instance of the paramount importance of this, it may be mentioned that many lives were lost when a large arch over the Derwent collapsed and fell into the river as the keystone was being placed in position. An allowance of one-third to one-half is usually allowed for waste on centring, or on conversion to use.

The removal of centres is generally accomplished by the striking of the folding wedges upon which they have been placed, or of the indented blocks which fit into corresponding blocks above and below, thus forming continuous wedges. The former is illustrated in fig. 112, and the latter in figs. 117 and 120; the wedges should be well rubbed with soft-soap and blacklead before being built upon. The French method is to destroy the ends of the principal rafters by degrees, but this cannot be done so evenly, and is a source of danger to the carpenter.

Perhaps the most ingenious method consists in having an iron cylinder filled to the extent of the upper half by a cylinder of wood, the lower half being sand; when it is required to strike the centres, holes, which are previously drilled in the base, are uncorked; the sand escapes and lowers the centre. It should always be remembered that centres should be struck slowly and evenly, so that the arch may gradually take its proper bearing.

CHAPTER X.

SCAFFOLDING, STAGING, AND GANTRIES.

SCAFFOLDING is a temporary structure placed alongside a building in order to facilitate its erection by supporting workmen, and raising materials during the construction, or for the repair of buildings.

Possessing a temporary character, the method of its construction is often likely to be overlooked by the student, but as a subject it is very interesting, and will repay him to give it the attention it deserves.

The various types of scaffolding, staging, &c., may be classified as follows :—

I. *Bricklayers' Scaffolds.*
II. *Masons' Scaffolds.*
III. *Staging.*
IV. *Gantries*—*a.* The gantry proper ; *b.* The gantry to support travellers.
V. *Derrick Cranes.*
VI. *Other Special Forms.*

I. BRICKLAYERS' SCAFFOLDS.—The ordinary bricklayers' scaffold is the one with which the student is most familiar, and fig. 122 shows such a scaffold, in isometric projection in order that it may be more easily understood. S are the *standards* or upright poles, generally of fir ; these are from 20 ft. to 50 ft. high, and from 5 in. to 8 in. in diameter at the lower or butt end. Smaller sizes—as, for example, those having a diameter of 2½ in.—are termed "rickers," and are generally to be obtained about 22 ft. long. These standards are usually fixed firmly in the ground about 4 ft. 6 in. from the face of the wall it is proposed to build, and are generally placed from about 10 ft. to 12 ft. apart.

In high building, where greater length is required, two standards are lashed together, butt to tip, and tightened up with wedges. In certain cases where it is inconvenient to place these standards beneath the ground—as, for instance, where it is not desired to disturb the pavement in a town—they are placed in tubs filled with earth, as shown to two of the standards in the illustration. To these standards, on the side nearest the wall, are lashed *ledgers* L, or horizontal runners, parallel to the wall, and from 3 ft. 6 in. to 5 ft. above each other; they are bound to the standards by rope, and

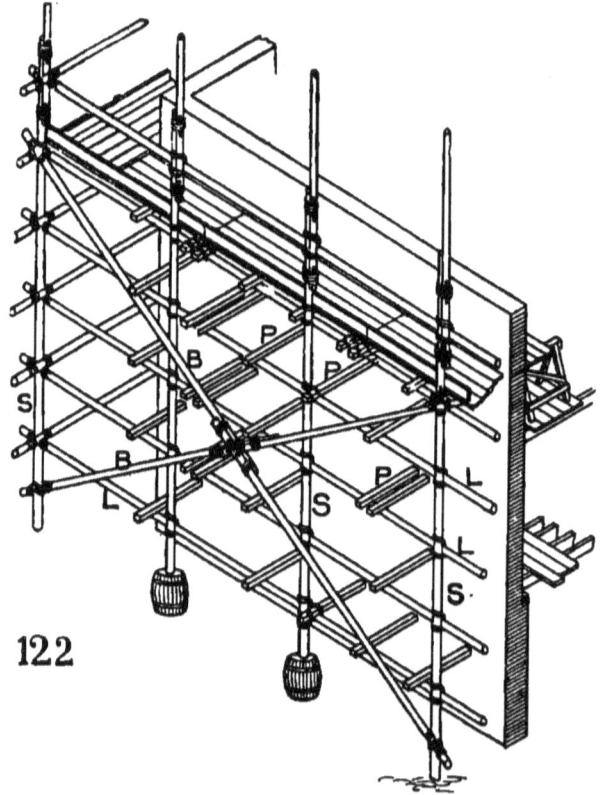

122. Bricklayers' Scaffold.

tightened by wooden keys or wedges introduced between the rope and the standards. The ledgers answer the double purpose of bracing the standards together at intervals in their height and of supporting cross pieces stretching from them to the wall. The latter are called *putlogs*, and are marked P on the illustration. They consist of squared imber, usually birch, above six feet long and of 4 in. by 3 in. scantling; they are placed about four feet apart, one end resting on the ledger and the other inserted in the wall by means of a hole left for the purpose by omitting a brick header. These putlogs support the scaffold boards, usually 9 in. by 1½ in. thick, with edges bound in hoop-iron. The scaffold boards are butted at their heading joints, the putlogs, where these occur, being placed only about 4 in. apart. On each staging, what are known as guard boards are placed as shown to prevent the materials or rubbish falling on to the ground below.

The scaffolding, as such, is now complete, but the whole is further stiffened and held together by long poles or *braces* B, lashed diagonally across every three or four standards. It will be understood that the scaffolding is raised as the building proceeds, and is therefore supported largely by the wall. When the wall has reached a certain height above the platform upon which the workmen are engaged, which is generally about 4 ft. 6 in. or 5 ft., as this is the greatest height that the average man can work with ease, another row of ledgers is lashed to the standards, fresh putlogs are inserted into the holes where brick headers have been omitted, and the scaffold boards are raised from the old to the new level. Meanwhile the ledgers and putlogs are left in position in order to steady the scaffold and are not taken out as a rule till the end of the operations.

It should be borne in mind that scaffolds of this description should not be unduly loaded, in such a manner as to press too heavily on the newly-executed work.

Fig. 130 shows the method to tying various knots used in scaffolding, which will be of interest to the student.

II. MASONS' SCAFFOLDS.—As the name implies, this form of scaffold is used by masons in the construction of stone ashlar walling, or indeed in any case where it is not possible or convenient to insert the ends of putlogs into the wall. Fig. 123 will sufficiently explain that this form consists of two frames parallel to each other, one about 4 ft. 6 in. from the face of the wall, as in a bricklayers' scaffold, and the other about 9 in. to a foot from the face of the wall to be built.

The object of this inner frame is to support the ends of the putlogs, as it is not convenient to leave holes in the face of a stone wall, of ashlar facing, for their insertion, as is done in a brick wall. Where openings occur in the wall, as at windows, advantage is taken to secure the frame to the wall itself in order to secure and steady the scaffolding; otherwise this form relies on its own rigidity, and is even

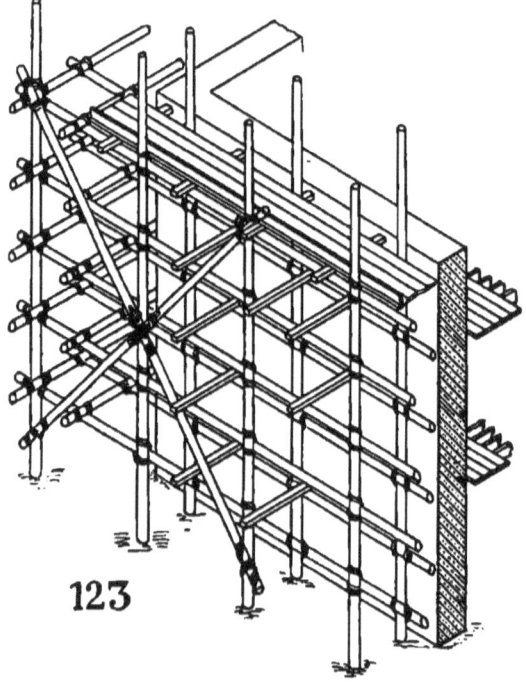

123. Masons' Scaffold.

called in some parts an independent scaffold. The general construction of such a scaffold should be stronger than in a bricklayers' scaffold, for the reason that heavier materials are used at one time. The standards are therefore placed closer together, and are more firmly braced. The most ordinary forms have now been touched upon, but in larger and more complicated structures, stronger and heavier erections have to be designed.

III. STAGING.—According to Tredgold, the term "gantry" is frequently applied to a structure which should be known as a staging, but he states that the former term should only be applied to a staging which is one story in height and carries a "traveller," and in this we shall follow him.

In buildings which have to go a certain height, it is evident that some type of construction heavier and stronger than ordinary scaffolding may have to be employed.

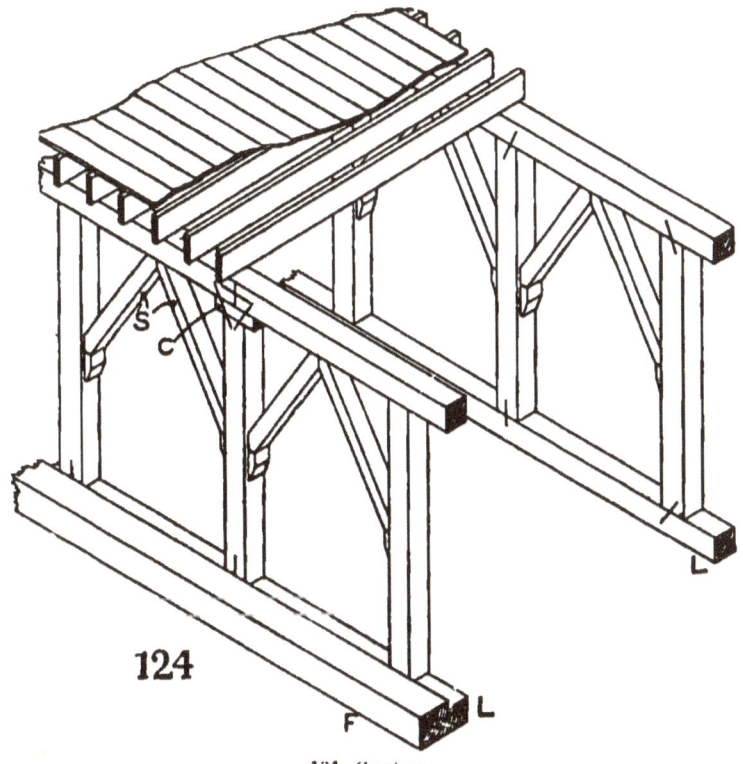

124. Gantry.

Such a construction, known as a *staging*, is shown in fig. 125. A beam of timber is placed across the head of each standard, and projects some 9 ft. or 10 ft. beyond it, at right angles to the direction of the runners on which it rests. This piece is called a "footing-piece," and serves the same purpose as the "foot-block" to a one-story gantry, as will be seen, except that it will be supported by the strut H. Such struts are usually in two pieces, in order that the strut F may pass between them.

The standards of the upper tiers should always be placed over those in the lower to prevent cross-strains in the hori-

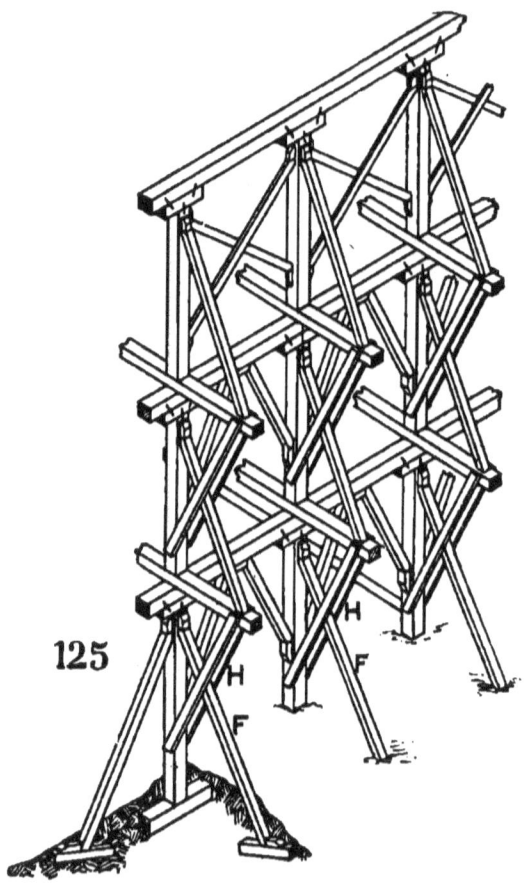

1.5. Staging.

zontal timbers, and diagonal bracing is frequently employed in the upper tiers, as shown. Staging has also to be largely employed in the construction of bridging and viaducts of great height. In such cases they are generally about the

eight of the springing of the arch, and are used to support
he requisite centring or as a platform for connecting the
ifferent sections of girders, &c. The example shown in
g. 126 exhibits the same principle as that adopted for
onstructing the land tubes of the Britannia Bridge in
850.

IV. GANTRIES.—Gantries are necessary where heavy
tones or other building materials have to be lifted, and in

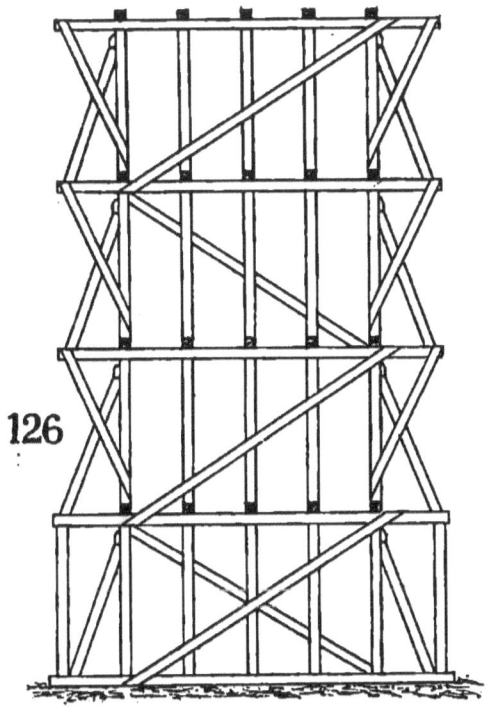

126. Staging used for the Britannia Bridge.

ases where ordinary scaffold poles would not be safe; they
re also used for supporting heavy machinery. Gantries
re much used in London and other large towns, as may be
oticed by the attentive student in his wanderings, and
uch can be learned by practical observation and sketching
uch examples as come before one. In spaces in front of
new building they are often placed over the public foot-

way and used as a store, or yard, from which the building operations are directed, and it is often on such elevated constructions that the clerk of works and foreman's offices are placed. (*a*) The gantry proper is shown in fig. 124, and is a structure commonly used and suitable for the uses named above. It will be seen that it consists of two frames formed of squared timber, about 10 ft. apart, and of scantlings from 6 in. to 12 in. square. The inner frame should be kept about 1 ft. from the face of the proposed wall. The frame is composed of *sleepers*, L, laid on to the ground, and protected from carriage traffic by a stout piece of timber called a *fender*, F. Upon the sleepers are placed the uprights secured by iron dogs. On the top of the uprights are placed pieces of timber or heads, which span from upright to upright, and are "dogged" in a similar manner to the sleepers. In order to distribute the pressure and decrease the bearing, the uprights are provided with a rough treatment of bracket capital as C. The weight on the whole structure is also distributed by struts S, usually about 5 in. by 5 in., pressing against each other at the upper end, and supported on cleats spiked to the uprights. The sectional area of these struts, in order to be effective, should not be less than half that of the uprights. The whole framework supports a platform, either constructed like an ordinary floor with joists and boarding, or by deals laid flat and touching each other. In these constructions the timbers should be weakened as little as possible, therefore the use of bolt-holes, notching, mortising, or otherwise cutting into the timber should be avoided. It is also necessary that the deterioration of the value of the timber should be reduced to a minimum. On account of both these reasons the several pieces are put together with dog-irons (fig. 105, chapter on "Shoring"), which are pieces of square or round iron about $\frac{3}{4}$ in. in diameter, having their ends pointed and turned down at right angles. They are driven well home, and can be removed with little injury to the wood. (*b*) A gantry to support a traveller is also affected by the conditions under which it is to be used. In this case the staging is not for the purpose of storage and manipulation of heavy material, but to provide a pair of rails along which a "traveller" may run. For this purpose the spacing is kept clear, the framing being connected at its ends; the intermediate portions being independent of each other. This is carried out, as shown in fig. 127, by keeping the struts outside the frames, the rest of the construction being similar to that already described. The lower end of the struts should always be fixed to foot blocks, as at G, by which they are prevented from sinking into the ground.

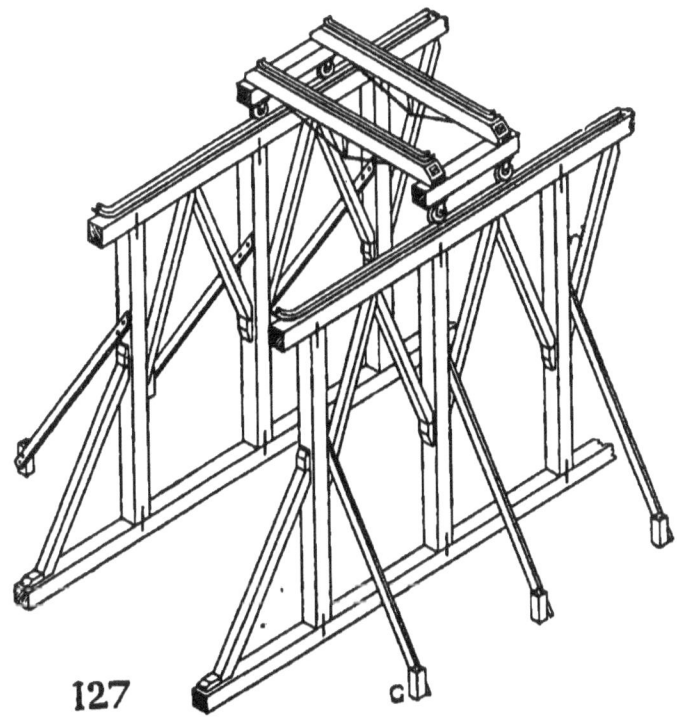

127. Gantry to support a "Traveller."

The traveller itself is constructed of two trussed beams placed at their extremities on two pieces of timber which rest on wheels running on rails extending the length of the framing. The mechanism of the traveller need not be touched upon here.

V. DERRICK CRANES.—Derrick cranes are being largely used in England at the present time, both on account of the diminution of labour which can be effected by their use and the time which can be saved. The student will not fail to observe those he may see in use, the following description is only intended as a general guide to the principles on which they are designed. Fig. 128 shows the timber towers for one of these derrick cranes. There are three in

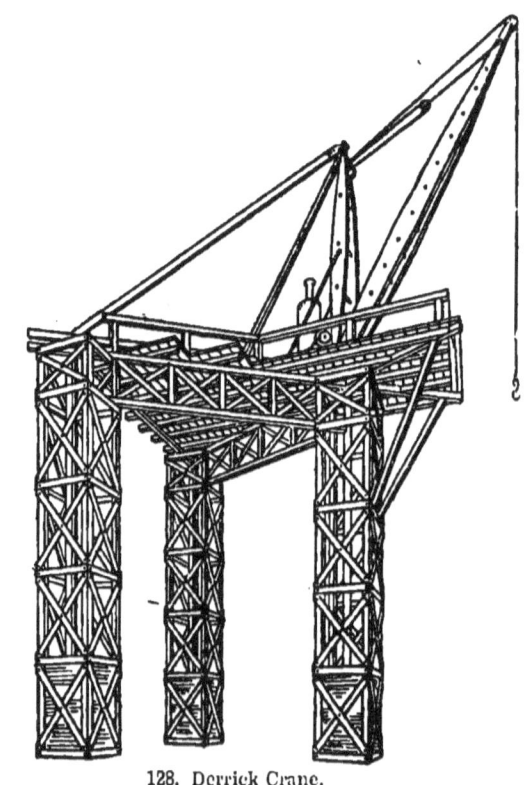

128. Derrick Crane.

number, distributed so as to form an equilateral triangle on plan; they are placed on platforms of wooden beams which serve as a base. The towers are usually about 6 ft. square, consisting of four uprights about 9 in. square, in one piece, or of deals bolted together, these are connected by cross pieces of, say, 9 in. by 3 in., about 6 ft. or 7 ft. vertically distant from each other, the bays thus formed are stiffened in the usual manner with cross braces 7 in. by 2 in. bolted to the uprights as shown. The bases of the towers are filled up with bricks or other heavy material to a weight at least double the load to be raised. The derrick is placed on one tower, which is connected with the other two at their upper ends by means of trussed beams as shown, and the two towers (other than the one on which the derrick rests) are connected by a single balk of timber strutted if necessary as shown. The derrick crane itself consists (see illustration) of sleepers, mast, jib, and stays, and it is anchored to the tower by means of chains which are passed over the sleepers on the top of the staging and connected to the platform at the base of each tower. On referring to the illustration (fig. 128) it will be seen that the mast, the upright member, may be formed out of one piece of timber or of pieces strutted apart and braced. It is placed on a pivot top and bottom, so that it may move in all directions. The jib supports the masses to be raised. It is attached to the mast at its lower end by a hinged joint, thus allowing freedom of action. At its outer end is a wheel, over which the chain to raise the material passes.

The angle of inclination of the jib can be altered by a chain in this portion. The sleepers lie on the stagings, and are connected to the lower ends of the mast by a swivel-joint. The stays are joined to the mast at its upper end by a swivel-joint, and to the sleepers at their lower end by a link-joint. These sleepers are anchored to the staging to keep the mast in position when the weights are put upon the derrick crane.

In some parts of the country, and especially on the Continent, buildings are erected without scaffolding, the work being performed from the inside, and the men supported on platforms raised on the floors of the building itself.

VI. OTHER SPECIAL FORMS.—Scaffoldings and stagings have to be designed for all purposes, and the architect should not be above advising on these points, as Sir Chas. Barry designed the scaffolds for the Houses of Parliament. Fig. 129 shows the revolving scaffold used for the repair of the dome of the Pantheon at Rome, and is an example of the skill of the Italians, who are especially clever at this kind of work. The scaffolds used in the erection of domes and roofs of considerable span consist of nothing more than a series of standards, with diagonal braces and struts.

The scaffoldings adopted in some of the recent large

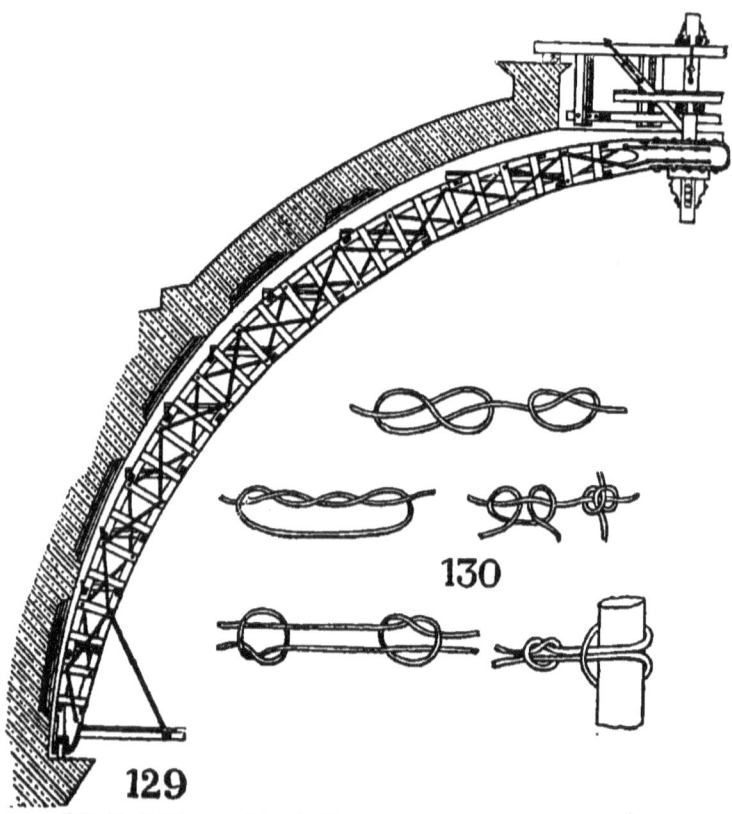

129. Scaffolding used in the Restoration of the Pantheon, Rome.
130. Various Forms of Knots.

exhibitions, such as Paris and Chicago, have been very ingeniously devised in regard to saving of labour and economy of material; but the subject of their construction can hardly be dealt with here. Especially in regard to the roofs were the scaffoldings worthy of study, the roofs being built in sections, the scaffolding was placed on wheels and moved along the length of the halls, as it was required; all the bays being erected by means of one section of scaffolding. Continental customs vary considerably from English methods, and in Greece the writer has seen scaffolding in which ladders are not used, each story being reached by inclined planes which the labourer can walk up. Each special case has to be designed to fit its purpose, and the student having grasped the main properties of scaffolding, can apply them with judgment to any particular problem.

CHAPTER XI.

PILLARS, BEAMS, AND GIRDERS.

PILLARS.—When a *pillar* or *column* is compressed in the direction of its length, the manner in which it behaves varies according to the ratio the length bears to its minimum diameter. If the length be great, the pillar will bend and fail by breaking at the centre as under a cross-strain; when however the pillar is very short, it will fail by crushing alone. It is generally assumed in practice that a pillar will fail by bending when its length exceeds thirty diameters, and it should not as a rule exceed twenty diameters in height. The formulæ generally used for pillars and columns are as follows when the length exceeds thirty times the diameter :—

D = diameter in inches.
L = length in feet.
W = safe load in cwts. (one-tenth breaking weight).
S = one side in inches.
B = breadth in inches.
T = the least thickness in inches.
E = 15·0 for teak.
 = 14·0 for English oak.
 = 12·0 for Baltic oak.
 = 12·0 for red pine.
 = 12·0 for ash.
 = 11·0 for Riga fir.
 = 11·0 for beech.
 = 9·0 for larch.
 = 8·0 for elm.
Then :—

For square columns .... $W = 1\cdot7 E \times \dfrac{S^4}{L^2}$

For rectangular columns .... $W = 1\cdot7 E \times \dfrac{BT^3}{L^2}$

For circular columns .... $W = E \dfrac{D^4}{L^2}$

For SHORT COLUMNS the above formulæ should not be used, but the calculation may be made by allowing for the safe resistance to compression per square inch of sectional area as follows :—
Memel or Dantzic fir ... ... 5 cwt.
English oak ... ... 6 cwt.

Beams.—At the present day wrought iron and steel have, to a large extent, taken the place of timber for spans of more than ordinary widths, owing to their greater strength, and also because they can be used of considerably less depth, thus saving the extra expense in the height of the building. The following formula will probably be found the simplest for calculating rectangular wooden beams supported at both ends, where—

BW = breaking weight at centre of beam in cwts.
B = breadth of beam in inches.
D = depth of beam in inches.
L = length of bearing in feet.
C = constant; 6 for teak, 5 for ash and oak, 4 for fir.

Then $BW = \dfrac{BD^2 C}{L}$,

In all cases where the load is uniformly distributed, double this weight will be required to cause fracture. If the load be applied at any other point, as at D, fig. 131,

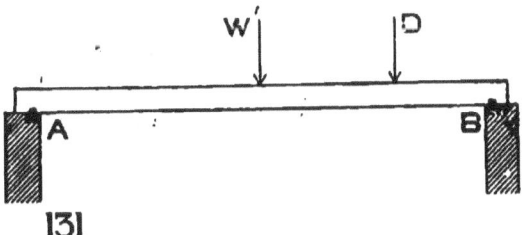

131. Simple wooden beam.

the breaking weight at that point will equal the square of half the length AB, multiplied by what the breaking weight would be if centrally loaded (W'); divided by the product of distances from the point of loading to each support.

For example, referring to the fig. 131, we obtain the following equation:—

$$BW \text{ (at the point D)} = \frac{(\frac{AB}{2})^2 \times W'}{AD \times DB}$$

Having found the breaking weight, the *safe load* may be found by dividing by 5, which is generally considered a sufficient factor of safety. From the above formula, and by substitution, any of the unknown items may be found, remembering that in *beams* the breadth is generally taken as two-thirds of the depth, and in joists as one-third.

Where the span is so great that it is found difficult to obtain the timber of sufficiently large scantling for the purpose, there are many methods for strengthening timber that can easily be applied, and these may be treated under the following heads:—

1. Flitched girders.
2. Built-up beams.
3. Girders trussed with wood.
4. Girders trussed with iron.
5. Framed truss girders.

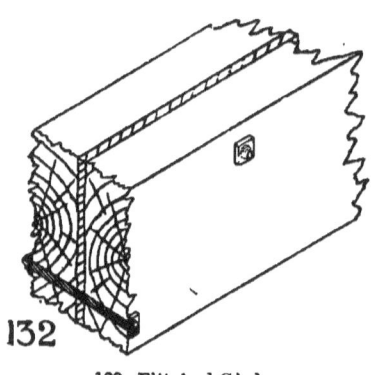

132. Flitched Girder.

1. FLITCHED GIRDERS (fig. 132) are formed by sawing a beam down the centre and bolting the two pieces back to back with an iron plate between them. It was customary when using large beams to do this without the iron plate, fillets being put between the halves to allow of the air circulating freely. This was a very good method as it gave an opportunity of examining the centre of the beam and reduced the timber to a smaller scantling, by which means it dried sooner and was less liable to rot; it was supposed by some to strengthen a girder, but it was really weakened by the

eration, though the method is good for the reasons stated
ove. The addition of the iron plate, which should not
 less in thickness than one-twelfth of the total width, con-
derably strengthens the girder. This was proved by
periment made at the Woolwich Arsenal some years
ick.
The following formula will be found useful for calculating
e strength of these girders, where—

BW = breaking weight at centre in cwt.
D = depth in inches.
L = bearing in feet.
B = breadth of wood in inches.
T = thickness of iron plate in inches.
C = constant; 4 for teak, 3 for oak, 2·5 for fir.

hen

$$BW = \frac{D^2}{L}(CB + 30T).$$

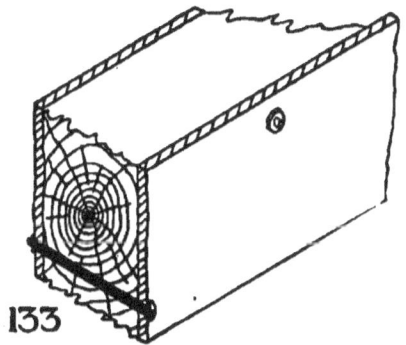

138. Flitched Girder.

he same formula may be used for a girder flitched as in
3. 133, which is a very
 eap and useful method
 strengthening beams
 situ, especially in the
 se of double floors, as
 obviates the necessity
 any considerable dis-
 irbance of the floor,
 hich would probably
 necessary if other
 eans were taken.
BUILT-UP BEAMS may
 either *coggled*, in-
 nted, or *curved*.
ig. 134 represents one of the former methods, which
 nsists simply in bolting two pieces together with oak
 ys in between to prevent the beams sliding upon one
 other. The grain of these keys should be at right angles
 that of the beams, and the depth of all the keys added
 gether should be between one-third and one-half more
 an the whole depth of the girder, and their breadth should

be twice their own individual depth. Sometimes the girder is cambered on its upper surface from the centre to the supports, and is held together by hoops instead of bolts, which are slipped on from the ends, and as they are driven closer to the centre they bring the two halves gradually together. When the girder is fixed if the joints are found to have given a little, the hoops may be driven still farther to the centre.

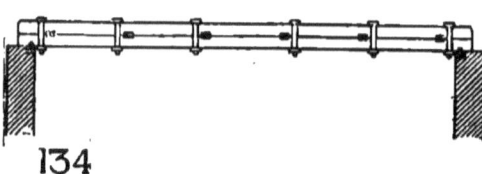

134. Joggled Beam.

Indented beams consist of indents instead of keys being used to prevent their sliding on one another. Fig. 135 represents an indented beam with the upper half in two pieces, so that a vertical king-bolt, tapering towards its ower extremity, can be inserted, which, by being screwed up on the underside, forces all the joints home. The depths of all the indents added together should not be less than two-thirds the whole depth of the girder. The upper half of these girders is sometimes constructed in oak or other hard wood; the lower half, being in tension, should be of tough straight-grained stuff. When there is more than one piece in the length of the upper half they may simply butt-joint against one another, but when this happens to be in the lower half, they must be scarfed, to resist tension as described in Chapter IV., pp. 32 and 33.

Curved beams (fig. 136) add considerably to the stiffness of girders, and Smeaton adopted this method for strengthening the beam of a steam engine. The upper

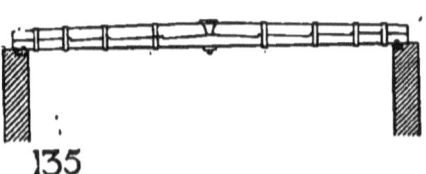

135. Indented Beam.

part is bent into a curve and prevented from springing back by straps or bolts, which should be very firmly secured resist any sliding of the parts. The thickness of the ent pieces should not be more than about one-fiftieth of the bearing, and the whole depth of the curved pieces should not exceed half the depth of the girder. It will be observed that in fig. 136 there is a butt-joint at

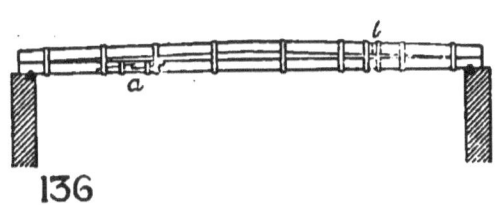

136. Curved Beam.

and a scarf at *a*; care should be taken that no joint in the lower part of the beam should be near the centre of the bearing.

GIRDERS TRUSSED WITH WOOD were formerly used to some extent, but they were found to be little, if any, stronger than when not trussed at all. Fig. 137 illustrates a fir girder trussed with oak. At first sight this looks very ingenious, but when it is remembered that all the additional strength that could be acquired would consist solely in the greater resistance to compression of the oak, which is slight, it will be seen that the gain is little, and unless the truss were very carefully made at the ends of the beam, it would probably be stronger without them.

GIRDERS TRUSSED WITH IRON may be divided into two classes ; (a) those trussed *within their own depth* and (b) those strengthened *with deep trusses*. In the former case they may be trussed with a *tension rod*, fig. 138. This is placed between the two halves of the balk, which has previously been cut longitudinally down the centre, and passes round a cast-iron bar placed centrally on the underside of the beam. The ends of the tension rod are secured with nuts, which press against cast-iron chairs at the extremities of the beam, or they may be screwed up

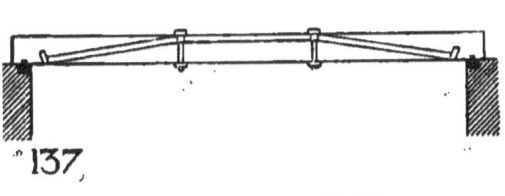

137. Girder trussed with Oak.

against a washer on the ends of the beam, previously cut off at right angles to the direction of the tension rod.

138. Girder trussed with iron tension rod.

This form of truss may be varied by the tension rods passing under two bars placed equidistant from the centre of the beam. Fairbairn's experiments on these beams proves that this method increases the strength of the beam about one-fifth, and he suggested that the rod should not be placed higher at the extremity of the beam than the horizontal line passing through the centre.

Another means of trussing girders in their own depth is shown in fig. 139, which consists of an iron king-bolt and struts, with a tension plate underneath, being inserted between the two halves of the beams. This form may be varied by having two queen-bolts instead of the king-bolt, the heads of the former being connected by an iron tie, and the struts being used as in the previous case.

The great objection to beams trussed in their own depth is that the ironwork becoming loose when the timber

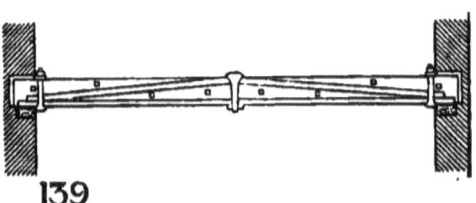

139. Girder trussed in its own depth.

shrinks, the whole weight is then thrown upon the timber which will be injured, unless the bolts are at once tightened up. After the girder is in position there is usually some difficulty in getting at the nuts for the purpose; owing,

140. Girder with deep truss.

therefore, to the adjustment of these beams becoming so easily disarranged they are not much used at the present time.

Girders with deep trusses may be used where the depth occupied will not be objected to. Fig. 140 shows a form of truss often used for beams to carry travellers, it is some-

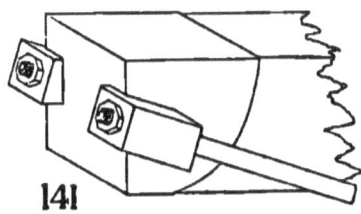

141. Iron Chair with ears to receive rods.

times strengthened by cross braces, as shown in dotted lines. In these types of trusses the rods, generally, are used on both sides, and are secured at the ends by ears on each side of the iron chairs (see fig. 141).

FRAMED TRUSS GIRDERS are nowadays much used in the construction of temporary buildings, notably in the

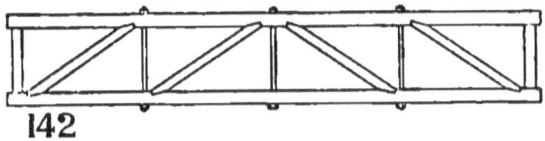

142. Framed trussed Girder.

structures of the Chicago Exhibition. Fig. 142 represents a very simple form connected by iron bolts. It must be remembered in designing a truss of this description that its strength does not altogether depend upon a balance of its parts, but the braces must be so disposed as to resist the pressure at the points where it will be applied. An illustration of this will be found by carefully studying the truss that carries the gallery floor, which will be found in the next chapter on floors.

The use of this kind of girder is of course very limited in ordinary buildings, owing to the depth required; but the transatlantic craftsman is far ahead of his British *confrère* in his ingenious adaption of these kinds of trussed girders in complicated and important structures.

CHAPTER XII.

FLOORS.

THE assemblage of timbers used for supporting the floor boards and ceiling of a room is called " naked flooring," and the construction of such timbers may be conveniently grouped under :—
1. Single-jointed floors.
2. Double floors.
3. Framed floors.
4. Composite floors.
5. Fire-resisting wood floors.
6. General remarks.

1. A SINGLE FLOOR consists, as its name implies, of one series of common or bridging joists. Fig. 143 shows a floor of this description in isometrical projection. Such a construction, it is affirmed, makes a much stronger floor with the same quantity of timber than a double or framed floor, but the great objection is that plaster ceilings are more liable to cracks and sudden jars from this form of floor, especially when used for long bearings. In order to make a stiff floor the joists should be thin and deep rather than thick and shallow; the least thickness which can well be employed is 2 in., as it is found that joists of a less thickness split when floor boards are nailed to them. In cases where joists cannot rest direct on the walls, as, for instance, where flues, fireplaces, or openings for staircases occur, a piece of timber called a *trimmer*, T, is framed between two of the nearest joists which pass such an obstruction and which have a bearing on the wall; these are strengthened for the purpose and called *trimming joists*, TJ. The intervening joists are tenoned into the trimmer.

The trimming joists must necessarily be made stronger in order to carry the extra weight thus put upon them, and in practice are usually made 1 in. thicker; although Tredgold's rule that $\frac{1}{8}$-in. should be added to the trimming joist for each joist supported by the trimmer would seem to be sufficient.

In order to prevent joists having a tendency to topple over or twist sideways, in cases where they exceed 8 ft. in span, a row of strutting should be introduced between them, this

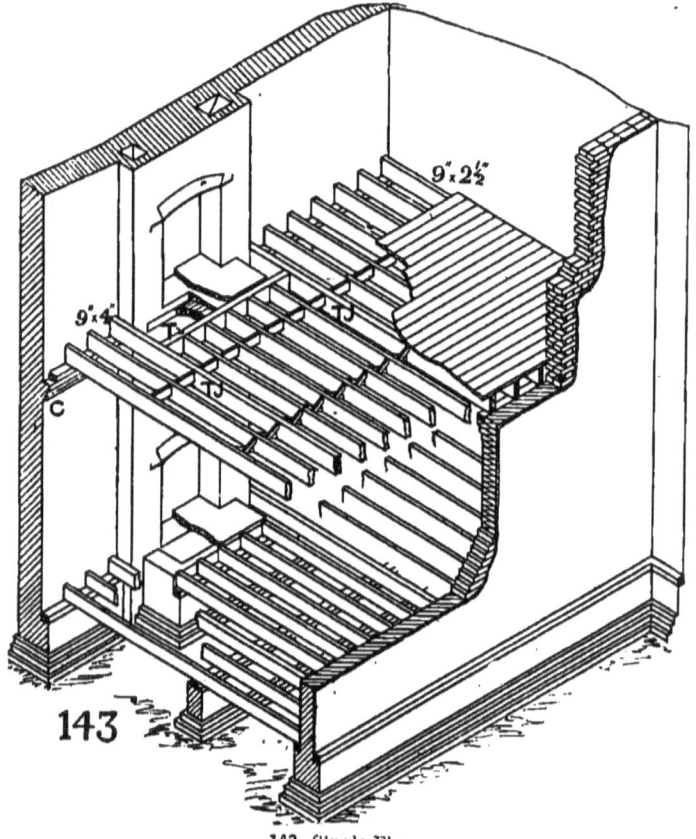

143. Single Floor.

also serves to stiffen the whole floor and equilibrate the pressure. When the bearing exceeds 12 ft., two rows of struts become necessary, another row being provided for every 4 ft. of bearing.

There are two methods of strutting in general use :—Pieces of joisting about the same depth as the joists themselves, and what is known as herring-bone strutting, formed of slips of wood, as shown, placed crossways and nailed between the joists. It is evident that single-joisted floors can be constructed to any length that joists can be obtained in one piece.

Tredgold recommends that 10 ft. should not be exceeded, as the strains produced by heavy furniture are liable to crack the ceiling below; but Tredgold always errs considerably on the side of strength, and in practice single-joisted floors are used in dwelling-houses for floors up to 15 ft. or 16 ft., which is the width of an ordinary dining-room.

In order to prevent the passage of sound from one story to another, pugging is sometimes resorted to. This consists in nailing rough fillets half-way down on each side of the joists, on these fillets are laid pieces of board which either support coarse plaster or concrete; as these, however, are liable to injure the joists by damp, a far better method is to place a layer of slag-wool (silicate of cotton) on the boards, as it is more effectively sound-proof, sufficiently light, and vermin-proof.

In ground floors the joists are generally supported on sleeper walls, and the span of the room being thus reduced the scantling of the joists can be considerably reduced. The lower part of fig. 143 shows a ground floor supported in this manner. The joists are laid direct on to wall-plates resting on these walls; where fireplaces occur it is usual to support the hearth on *fender-walls*, as shown, which then enables the hearth to be supported in a solid manner, and does not require the trimming necessary for upper floors.

Tredgold gives these as being spaced 12 in. apart from centre to centre, but they may be, and in fact are usually, spaced 12 in. apart in the clear. Although the scantlings are given up to 20 ft., it is seldom that this type of flooring is used for more than 15 ft. span.

Tredgold formula for the scantlings of common joists is as follows, where L = length in feet; B = breadth in inches; D = depth in inches.

$$D = 3 \sqrt{\frac{L^2}{B}} \times 2\cdot 2 \text{ for fir, or } 2\cdot 3 \text{ for oak.}$$

2. DOUBLE FLOORS are used for spans over 15 ft. or 16 ft., the object being to reduce the bearing of the common or bridging joist. For this reason *binders* (fig. 144) are placed

across the room to be floored at distances from 6 ft. to 10 ft., and the bridging joists are laid across these as shown in the illustration.

As the whole weight of the floor comes upon these binding joists, care should be taken that they are not placed over openings, but upon the piers between such openings. The distance between the binders is of course regulated by the distribution of solids and voids, and must be carefully con-

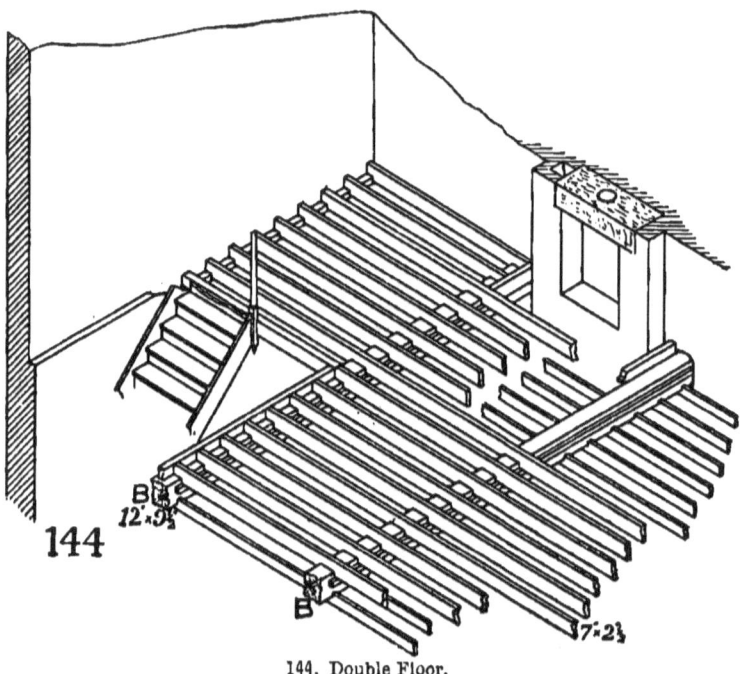

144. Double Floor.

sidered by the architect. For this reason a double floor is often more satisfactory than a single floor, as it brings the pressure on to points calculated to receive it. The binders should rest on stone templates placed in a recess in the wall, thus allowing of an air-space all round the ends of the binder, and preventing decay or dry rot. The common joists are either notched as shown in fig. 144, and supported on fillets nailed to the side of the binder, or the binder is notched and the common joists laid across. As it is necessary to weaken the binder as little as possible, the former is the better method. In double floors the ceiling may be placed on the underside of the common joists, and the binder treated so as to be visible below; or ceiling-joists are placed from binder to binder, and fixed in some cases by means of a chase-mortise, as shown in fig. 43 in a previous chapter. One advantage of this form of floor is that the passage of sound from one room to another is reduced to a minimum by the air-space formed between the ceiling-joists and bridging

A Table of Scantlings for Bridging Joists is here given, being revised from Tredgold:—

Bearing in feet.	Breadth, 1½ in.	Breadth, 2 in.	Breadth, 2½ in.	Breadth, 3 in	Breadth, 3½ in.	Breadth, 4 in.
	Depth in inches.	Depth in inches.	Depth in inches.	Dep inch	Depth in inches.	Depth in inches.
5	5¾	5¼	4¾	4⅔	4⅛	4
6	6¼	5¾	5⅜	5	4¾	4½
7	7	6½	6	5½	5½	5
8	7¾	7	6½	6¼	5⅞	5¾
9	8½	7½	6⅞	6½	6	5⅞
10	9	8	7½	7	6¾	6½
11	9½	8¾	8	7½	7	6¾
12	10	9½	8½	8	7½	7¼
13	10¾	9¾	9	8½	8	7¾
14	11	10	9½	9	8¼	8
15	11½	10½	9¾	9¼	8¾	8½
16	12¼	11	10½	9¾	9¼	8⅞
17	12¾	11½	10¾	10¼	9¾	9¼
18	13¼	12	11¼	10½	10	9½
19	13¾	12½	11½	10¾	10¼	10
20	14¼	13	12	11¼	10¾	10¼

joists; the latter are stiffened by having a bearing of not more than 10 ft., and the ceilings in consequence are not so liable to crack from vibration.

A table of scantlings for binding joists is given on page 125 Tredgold's rule for scantlings of binders, 6 ft. apart, where D = depth in inches, B = breadth in inches, L = length in feet.

$$D = 3\sqrt{\frac{L^2}{B}} \times 3\cdot 42 \text{ for fir, or} \times 3\cdot 53 \text{ for oak.}$$

$$B = \frac{L^2}{D^2} \times 40 \text{ for fir, or} \times 44 \text{ for oak.}$$

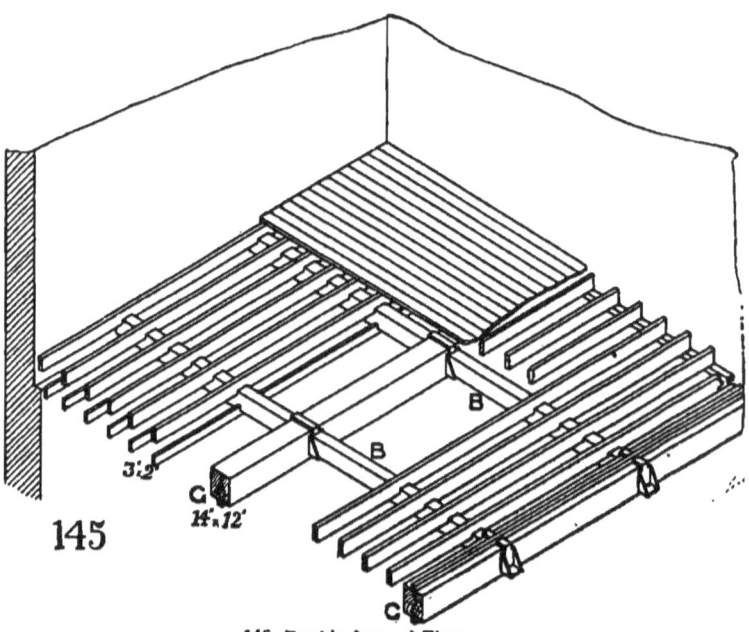

145. Double-framed Floor.

3. FRAMED FLOORS.—This type of floor (fig. 145) is used for spans of large dimensions, and consists of girders, binders, and bridging joists. The girders are usually placed about 10 ft. apart from centre to centre, or as dictated by the exigencies of the plan.

The following is for girders of Baltic yellow pine 10 ft. from centre to centre.

BINDING JOISTS OF BALTIC YELLOW FIR. DISTANCE FROM CENTRE TO CENTRE, 6 FT.

Bearing in feet.	Depth, 6 in.	Depth, 7 in.	Depth, 8 in.	Depth, 9 in.	Depth, 10 in.	Depth, 11 in.	Depth, 12 in.
	Breadth in inches.	Breadth in inches.	Breadth in inches.	Breadth in inches.	Breadth in inches.	Breadth in inches.	Breadth in inches.
5	4¾	3	2	—	—	—	—
6	6¼	4	3	2	—	—	—
7	—	5½	4	2¾	2	—	—
8	—	7	5¼	3¼	2¾	2	—
9	—	—	6	4½	3½	2½	—
10	—	8	5½	4	3	2½	—
	Depth, 13 in.	Depth, 14 in.	Depth, 15 in.				
11				6¾	5	3¾	3
12	—	—	—	8	6	4½	3½
13	Breadth in inches.	Breadth in inches.	Breadth in inches.	—	7	5¼	4
14				—	8	5¼	4½
15	4	3¼	2¾	—	9	6¾	5¼
16	4½	3¾	3¼	—	10¼	7¾	6
17	5¼	4¼	3½	—	—	8¾	6¾
18	5¾	4¾	4	—	—	10	7½
19	6⅓	5¼	4¼	—	—	—	8½
20	7¼	6	4¾	—	—	—	9½

The girders either have the binders, B, framed into them by means of a short tenon known as a "stub-tenon," or—what is more frequently adopted nowadays—an iron stirrup is placed over the girder, as shown in the figure, and in this stirrup the binder rests. This is a much better method, as the girder is not weakened in any way by being cut into to form the mortise, and it should always be employed in a bearing of any size, in preference to the old-fashioned mortise and tenon. When, however, the mortise and tenon are used, the joints of two opposing binders should not be placed opposite each other, as this would weaken the girder too much, but should be placed in intermediate positions.

Tredgold's rule for scantlings of girders is—

$$D = 3 \sqrt{\frac{L^2}{B}} \times 4\cdot2 \text{ for fir, or} \times 4\cdot34 \text{ for oak.}$$

$$B = \frac{L^2}{D^3} \times 74 \text{ for fir, or} \times 82 \text{ for oak.}$$

4. COMPOSITE FLOORS.—By the increasing use of iron and steel in buildings, it has become usual to employ girders of these materials in spans over 16 ft., especially in towns where they are easily procurable; and as the carpenter has to fix his joisting to these girders, it has been thought advisable to consider these under the heading of composite floors. Fig. 147 shows their general setting out, in which it will be seen that the common or bridging joists are fixed or spiked to wood plates, which are bolted to the iron joists. This method of construction is being extensively used, and may be said to have superseded the old timber forms, where iron is attainable.

The strengthening and trussing of girders has been taken under a special chapter, specially devoted to that subject, and need not here be referred to.

5. FIRE-RESISTING WOOD FLOORS.—A form of floor which has been used occasionally, and which has much to recommend it, is that shown on page 130 (fig. 148). The patent was originally taken out by Messrs. Evans and Swain. It will be seen that the joists are laid alongside,

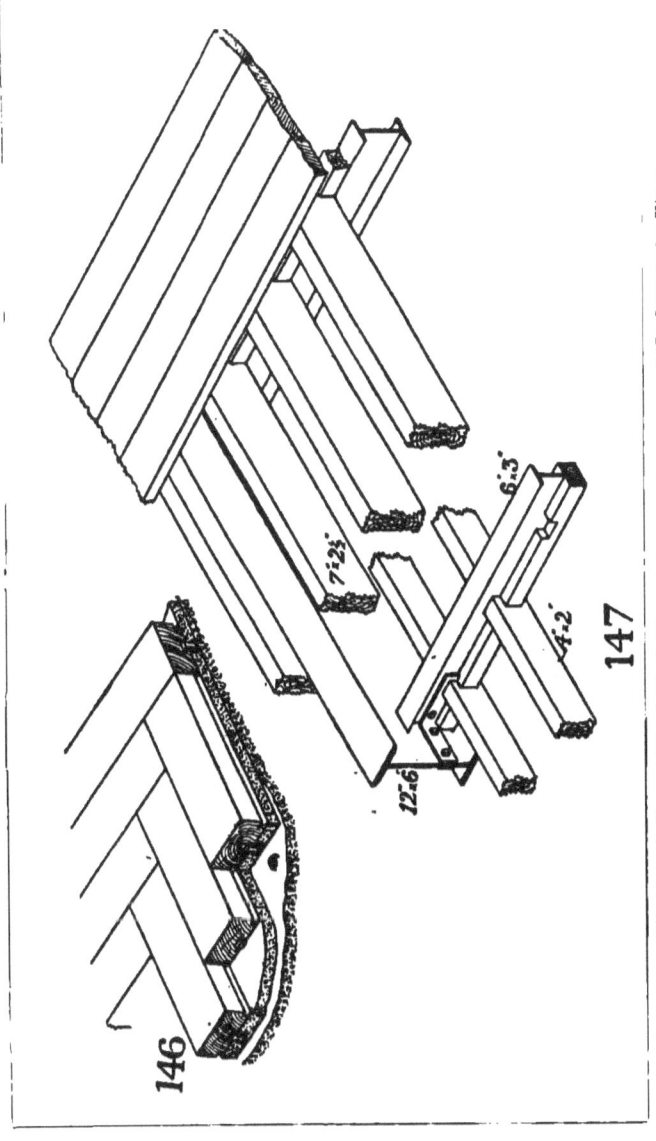

146. Geary's Patent Wood-block Floor. 147. Composite Floor.

touching one another. The depths of the joists in a construction of this kind varies from 4½ in. in a span of 8 ft. to 11 in. in a span of 30 ft. The under-side may be planed and left visible, or provided with a plaster ceiling, in which case the key is obtained by counter-lathing, as shown. This system of construction is largely in use in the New England States of America, and is employed there for the floors of warehouses.

In one example which came under the writer's notice the floors were constructed of 9 in. by 3 in. joists, touching each other; over these was placed a layer of asbestos paper and

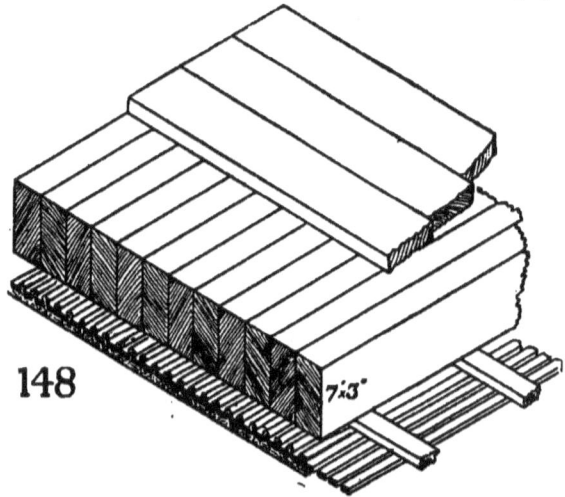

148. Fire-resisting Floor.

two layers of floor boarding, with their joints running in the opposite direction. The joists were supported on wooden beams resting on oak uprights, which were 2 ft. 6 in. square in the basement.

As showing the belief in this system as a fire-resisting construction, the American insurance companies give preferential rates for buildings of this class.

6. GENERAL REMARKS.—Girders should never, if possible, be laid over openings, but where it is unavoidable, the plates should be strong enough, or should be supported sufficiently

Bearing in feet.	Depth 10 in.	Depth 11 in.	Depth 12 in.	Depth 13 in.	Depth 14 in.	Depth 15 in.	Depth 16 in.	Depth 17 in.	Depth 18 in.
	Breadth in inches.	Breadth in inches.	Breadth in inches.	Breadth in inches.	Breadth in inches.	Breadth in inches.	Breadth in inches.	Breadth in inches.	Breadth in inches.
10	7½	5½	4	3½	2¾	2¼	1¾	1½	1¼
11	9	6¾	5	4¼	3¼	2¾	2¼	1¾	1¾
12	10½	8	5¾	5	3¾	3¼	2¾	2¼	2
13	12½	9½	6¼	5¾	4½	3¾	3	2½	2¼
14	14½	10¾	7¾	6¾	5¼	4½	3½	3	2½
15	16½	12½	9	7¾	6	5	4	3½	3
16	18½	14	10¼	8¾	7	5¾	4¾	4	3¼
17	—	16	11¼	9¾	8	6¼	5¼	4½	3¾
18	—	17¾	13	11	8¾	7¼	6	5	4
19	—	19¾	14½	12	9¼	8	6½	5½	4¾
20	—	—	16	13½	10¾	9	7¼	6	5¼
21	—	—	17½	15	12	9¾	8	6¾	5¾
22	—	—	19¾	16½	13	10¾	8¾	7½	6½
23	Depth	Depth	Depth	18	14¼	11½	9½	8	6¾
24	19 in.	20 in.	21 in.	19½	15¾	12¾	10½	8¾	7½
25	—	—	—	—	16¾	13¾	14½	9½	8
26	Breadth	Breadth	Breadth	—	18¼	15	12¼	10¼	8½
27	in	in	in	—	19¾	16	13	11	9¼
28	inches.	inches.	inches.	—	—	17¼	14¼	11¾	10
29	—	—	—	—	—	18½	15¼	12¼	10¾
30	9¾	8½	7¼	—	—	19¾	16¼	13½	11½
31	10¼	9	7¾	—	—	—	17¾	14½	12¼
32	11	9½	8¼	—	—	—	18½	15½	11
33	11¾	10	8¾	—	—	—	19¼	16½	14
34	12½	10¾	9½	—	—	—	—	17½	14¾
35	13¼	11½	10	—	—	—	—	18½	16¾
36	14	12	10½	—	—	—	—	19½	16½

to place the weight on the piers. *Ceiling-joists* for double and framed floors are generally of stuff about 3 ft. by 2 ft., or some such scantling as will cut out of a deal without waste. They should be supported at least every eight or ten feet. In single floors they are frequently dispensed with, the lathing and plastering being fixed to the under-side of the floor-joists. They should not be placed further apart than 14 in. centre to centre, and in single floors, every fifth or sixth bridging joist is often made 2 in. deeper than the other joists, and the ceiling-joists are affixed to these.

Wall-plates should, of course, be made stronger as the span becomes longer, and the following scantlings are recommended :—

For a 20-ft. bearing, wall-plates 4½ by 3.
,, 30-ft. ,, ,, 6 by 4.
,, 40-ft. ,, ,, 7½ by 5.

When not of sufficient length, they are connected by means of halving, bevelled halving, or dovetailed notching; these joints are shown in the article on joints in carpentry. The joists are either spiked on to the wall-plates, which is the most usual way, or they are notched and nailed. Wall-plates may either rest on offsets from the wall, as in the ground story (fig. 143) or they may rest on corbels of brickwork made to receive them, as shown at C, either of which is much better than the usual method of resting them on the inner part of the wall, thus interfering with the general thickness of the wall. In alterations or additions, or in rebuilding where party-walls already exist, a good plan is to insert heavy angle-irons at intervals, into which the wall-plates rest.

TRIMMING.—It has been explained that where flues occur, it is necessary, in order to prevent the ends of joists entering them, to trim round these obstructions. Trimming has also to be executed for openings in floors, as where staircases (fig. 144) or trap-doors occur. The illustration (fig. 143) shows the method adopted in order to avoid a fireplace, and the method of forming the trimmer arch to support the stone hearth. According to the London Building Act, 1894, the hearthstone has to be 12 in. longer than the width of the chimney-opening, and to be at least 18 in. in front of the breast, so this regulates the size of the space to be trimmed. The brick-trimmer arch is thrown from the brickwork of the breast to the "trimmer," and sometimes it is supported by means of a fillet formed as a skew-back to the side of the trimmer. In cases where there are no ceiling-joists, what are known as "filling-in" pieces are inserted to support the laths for the plastering of the ceiling. The method shown of forming a trimmer arch refers to a case in which the joists run at right angles to the wall; when they are parallel the trimmer arch is turned against one of the continuous joists which becomes the trimming joist, and

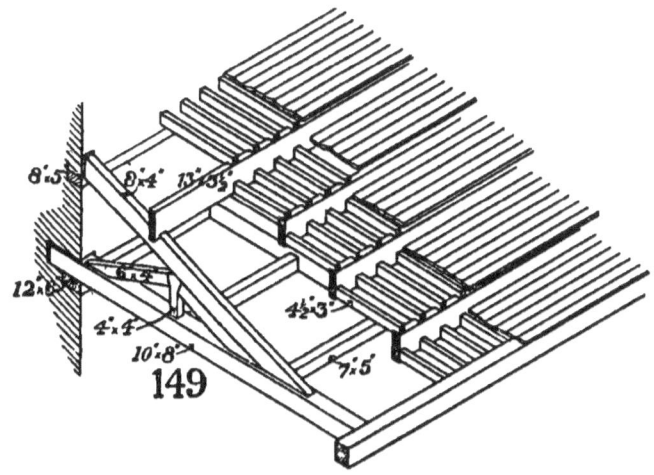

149. Gallery Floor.

short trimmers are carried from this to the wall. It may be mentioned that trimmer arches are being less used, their place being taken by coke breeze concrete, which is supported on fillets nailed to the joists surrounding the opening.

It has already been intimated that timbers should have a clear space all round their ends to prevent decay; in no case should they be allowed to be built into the wall. In addition, all floors should be thoroughly ventilated by means of air-bricks in the outer walls so as to ensure a thorough current of air. For the same reason care should be taken to leave holes in the sleeper walls. For the sake of economy joists should be placed the narrowest way of the room and should be used so as to form a tie to the enclosing walls. This is a very important point and contributes largely to the stability of the building. For this reason joists may, with advantage, be made to go in different directions over each floor to form horizontal ties at points in the height of the building. Those who have travelled on the Continent, especially in Belgium and Holland, will have noticed the charmingly designed iron ties on the face of the brickwork, these are connected with the floors by means of bolts and therefore conduce to the stability of the wall.

7. GALLERY FLOORS.—Gallery floors may be either held upon cantilever trusses or supported in some such way as shown in our illustration (fig. 149), which is one of the trusses of the side gallery of a chapel. The framing should be strong and secure, as galleries are frequently crowded with a large number of people. The illustration is merely given as an example, each case has to be specially provided for. The sizes are in some cases governed by rules, as children's galleries in Board Schools under the Education Department. The scantlings are marked on the example given.

Having touched on the various forms of *naked flooring* in general use, we may now proceed fitly to discuss floor-coverings.

CHAPTER XIII.

FLOOR-COVERINGS.

HAVING treated of the various forms of naked flooring, the subject which naturally follows is the covering with which they are overlaid and its method of fixing, from the carpenter's or joiner's standpoint. The subject can be conveniently treated under the following headings :—

1. General remarks.
2. Ordinary flooring.
3. Rebated and other forms.
4. Wood-block floors.
5. Parquet floors.
6. Special floors.

1. GENERAL REMARKS.—Floors may be either laid *straight joint*, in which the side joints of the boards are continuous throughout their length, which is the most usual way, or they may be laid *folded*, that is, when the side vertical joints are not continuous, but are set out in bays of four or five boards in one *bay* or *fold*. This latter method is more often adopted on the Continent.

The boards used for flooring are battens or deals, which should be perfectly free from sap, large loose or dead knots. As soon as possible after the building has been commenced and is ready for their reception, they should be laid out in the building across the floor joists, bottom upwards, so that they may have every opportunity of drying and seasoning without being damaged by being walked upon. In laying floor boards, they may be either nailed down one after another, or in cases where it is expected they are not properly seasoned, one is laid, then a fourth, leaving a little less space for the two intermediate ones than their actual width; these are then forced into position by workmen jumping on the two intermediate boards, and thus forcing them into position, when they are permanently fixed by brads. In this method the boards are said to be folded.

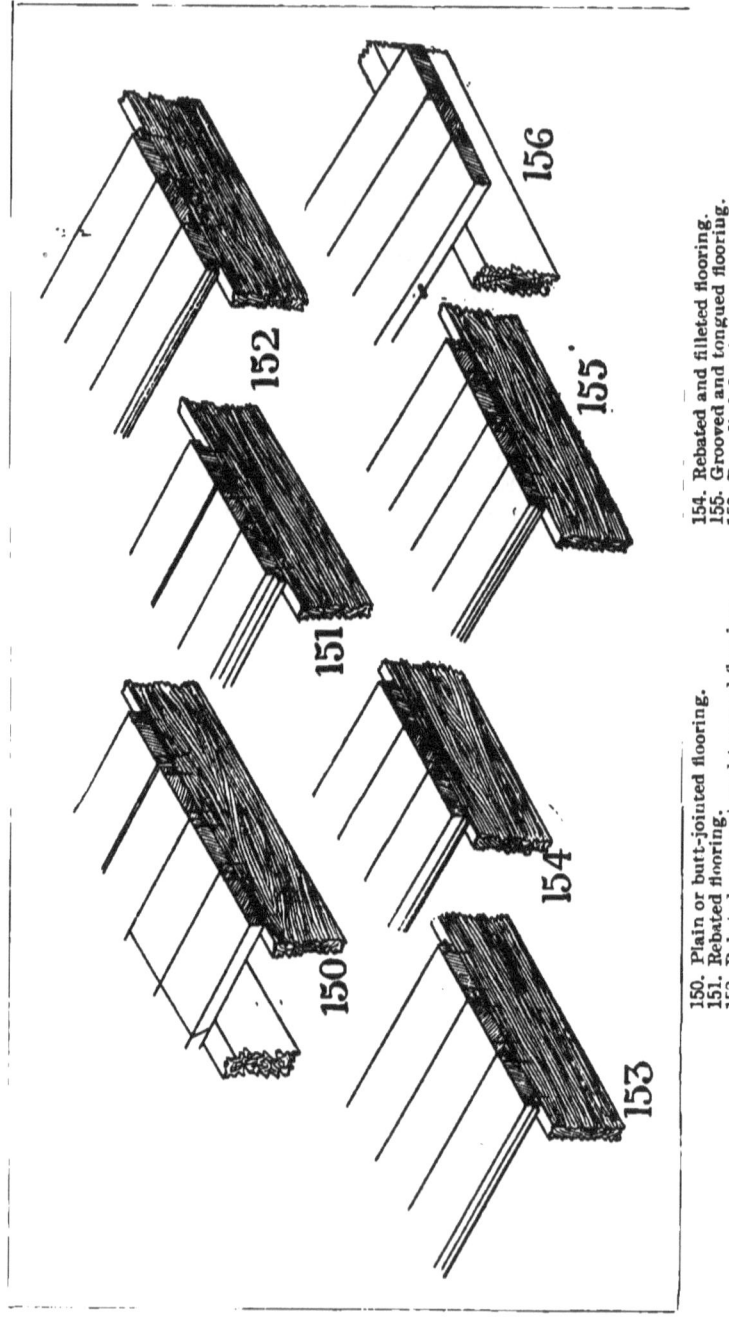

150. Plain or butt-jointed flooring.
151. Rebated flooring.
152. Rebated, grooved, and tongued flooring.
153. Ploughed and tongued flooring.
154. Rebated and filleted flooring.
155. Grooved and tongued flooring.
156. Dowelled flooring.

In order to minimise the effects of shrinkage floor-boards should be laid in as narrow widths as possible, from batten widths to strips of 3 in. and 4 in., as the joints can be kept tighter. The edge of each joint is "shot" or smoothed off with a plane, forming what is known as a butt or plain joint. The boards should all be brought to the same width, have their edges shot, and be gauged with a fillister-plane. In laying they are generally brought tightly together by means of flooring cramps; the method of being "laid folding," already described, is only used in common floors. The board, being cramped into position, is secured to the joist by "flooring brads," or flat-sided nails, which, being driven in parallel to the grain, have no tendency to split the board. Secret nailing is adopted in rebated floors, and in the better class generally, and is seen in fig. 152, which sufficiently explains the process. The boards being seldom long enough to go right across the room their ends have to be butted against an adjoining board; this is called a *heading* joint, and it is evident that these should occur on a joist in order to make a satisfactory junction, and that they should break joint on plan in order to counteract any tendency of the boards to slide out of position. There are different ways of forming these heading joints. They may be simply *butt* joints, or, as is usually the case, they may have splayed or bevelled headings, as at H in fig. 150. This is the better course, as it is more likely to keep the ends in position. Other forms are grooved and tongued, rebated and tongued, or forked headings, which are seldom used in practice.

2. PLAIN OR BUTT-JOINTED FLOORING.—Fig. 150. In this the boards are simply laid side by side, their edges having been previously shot, and are held down to each joist by two nails.

3. REBATED AND OTHER FORMS.—The practice of joining the edges of boards by means of rebates, or of tongues and grooves, appears to have been introduced into England in the fifteenth century. The inevitable shrinkage in floors will always cause a straight-jointed floor to open along the edge. This is to be avoided in all cases, but it is especially harmful in certain cases. For instance, in a ground floor resting on sleeper-walls, and having no ceiling beneath, the

K

air would blow up through the joints and cause draughts, and in a warehouse, where no plaster ceilings are put, the dirt and dust would pass from one floor to another unless some means, such as we shall show, were introduced as a preventive. It is advisable, therefore, for sanitary reasons, to use some form of tongued floor. Other reasons, such as the unsightliness of the nail-holes, prove the necessity for the rebated floor. There are several forms, some of which we illustrate.

Rebated flooring.—This is shown in fig. 151, and consists in each board having a rebate on each side, fitting into that of the adjacent one, the boards being nailed as in an ordinary floor.

Rebated, grooved, and tongued.—Fig. 152 is an extension of the above joint, in which it will be observed that besides the rebate a groove and tongue is worked on the adjacent edges of each board. In this case, the boards can be secret-nailed, as shown, and they are held in position by the tongue. This form is used in good floors, in which visible nails would be an eyesore.

Ploughed and tongued.—Fig. 153. In this form, a groove is run along the side of each board by means of a plough-plane, and a wooden or iron tongue is inserted; these should be kept as low down as possible, in order that by the wear of the floor it may not be reached. This form is specially suitable for warehouses, in which no plaster ceilings are fixed, as it prevents dust from descending through the open joints of the floor, and is not much more expensive than plain shot edges.

Grooved and tongued.—Fig. 155 shows this method, which is not often used. A groove is run along one board, and a tongue on the adjoining board fits into it. This form possesses the disadvantages of cost, and has no advantages over the last-named joint.

Rebated and filleted.—Fig. 154. This form is essentially one for floors on which rough work is anticipated, as in warehouses and barracks. A thin rebate is taken out of the underside of each edge of the board, which is occupied by a metal or wooden tongue. It will be seen that almost the whole depth of the floor is available for wear before the tongue is exposed.

Dowelled.—Fig. 156. Dowelled floors have small oak dowels fitted at intervals along the edges of the boards, into holes formed to receive them. They are little used, and do not possess any advantage over a good ploughed and tongued floor.

4. WOOD-BLOCK FLOORS.—This form of covering has of late years been increasingly used. It is formed of blocks of wood varying from about 9 in. to 12 in. in length, 3 in. in width, and 1 in. to $2\frac{1}{2}$ in. in depth. These are usually laid on a solid bed of concrete, on the solid ground, or in conjunction with some fireproof system of construction in upper floors. The advantages are that it leaves no space for vermin, is free from the noise and echo of common boarded floors, and is easily replaced. There are several varieties of

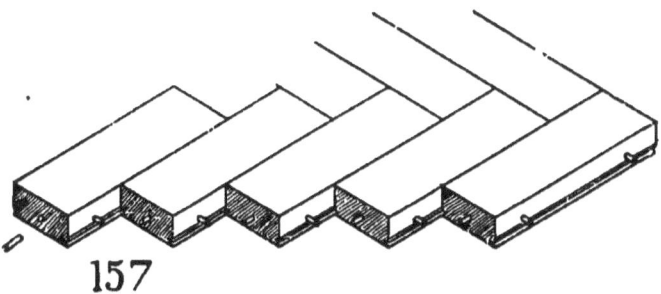

157. Duffy's Patent Wood-block Flooring.

wood blocks, and many patents have been secured in regard to their laying. These all have the same object in view, namely, to counteract any tendency the blocks may have to rise, either from damp or other causes. Geary's patent invincible system (fig. 146) has every block secured to the substructure by means of metal keys or studs dovetailed into the blocks, one end of the key being firmly embedded in the mastic or bitumen on which the blocks are laid. Duffy's patent wood-block flooring (fig. 157) has been much used. In this patent the wood blocks have dowels placed, as shown, in holes in each block, thus keeping each in its place. Security is further obtained by the lower edges being dovetailed and laid in a bed of bitumen composition. This

K 2

flooring is well adapted to all kinds of work. For churches, wood blocks $1\frac{1}{2}$ in. may be taken as sufficient; for schools, $1\frac{3}{4}$ in. to 2 in.; and for factories and barracks, where very rough wear is expected, blocks as thick as $2\frac{1}{2}$ in. may be employed.

5. PARQUET FLOORS are composed of thin layers of wood worked into some geometrical pattern, either extending over the whole room or used only as a border round the room. The parquet may be laid direct on to the joists or is often laid on the top of an ordinary floor. This type of flooring is kept clean by the use of turpentine and waxing, and is in every respect a more sanitary finish to a room than ordinary washed floors. Parquet is made in all thicknesses from $\frac{1}{4}$ in. to 1 in., and of oak, walnut, or teak, &c.

Parquet floors are also made to be removable, and are often laid down for temporary purposes, as for dances, &c. Mackenzie's patent removable parquet flooring is easily fixed, each section being grooved and tongued with metal tongues, and fixed with secret nails. This flooring has a metallic backing which prevents its being affected by any shrinking of the sub-floor. Cork-carpet floors are also frequently laid down for bathrooms, as the material being a non-conductor it is warm to the touch.

There are various other patents for these floors, having the same object in view, but they need not be mentioned here.

6. SPECIAL FLOORS.—Floors have often to be constructed for special purposes. Those used for dancing should always be specially treated, so as to give the necessary spring. To effect this object floor-boards are often laid on indiarubber, as at the King's Hall, Holborn, and elsewhere. At the King's Hall the flooring to the main hall is floored with oak in 3-in. widths, laid on strips of indiarubber placed on the joists, about $\frac{1}{2}$ in. thick, in order to give the requisite elasticity for dancing purposes.

Rondelet, in "L'Art de Bâtir," mentions a peculiar floor erected at Amsterdam, in which no joists whatever are used. It is a room 60 ft. square. Strong wall-plates are fixed round the room, secured with iron straps at the angles, and rebated to receive the flooring, which consists of three thicknesses of 1½ in. boards without any joists. The first layer is placed diagonally across the room, and made to rise 2½ in. higher in the centre than the sides. The second layer of 1½ in. boards is also laid diagonally, but the reverse way to the first, to which it is well nailed. The third layer of 1½ in. boards is placed parallel to one side of the room, and is also well nailed to the boards beneath. All the boards are grooved and tongued to each other, and thus form a solid floor 4½ in. in thickness. Other instances could be named, which are not, however, of interest in a practical way.

CHAPTER XIV.

FRAMING IN PARTITIONS AND FRAME HOUSES.

BEFORE treating of the subjects of partitions and framework houses, a few remarks on the designing and jointing of trusses may be of service and interest. The object that should be sought in a system of framing is to direct the pressure into the longitudinal direction of the timbers composing the frame ; therefore the joints should be constructed so as to direct the pressure down the axes of the pieces. It often happens that, by settlement or shrinkage, the pressure has to be sustained entirely upon the angular points of the joints, which are injured by the strain, and thus often cause a further settlement. In timber structures of magnitude this becomes manifest to an alarming extent, a fact that must have been brought home to Perronet, when seven or eight pieces of each frame of his centre for the bridge at Neuilly (see chapter on "Centres," fig. 116) were fractured from end to end, even although the joints were not very oblique.

When one piece of timber is perpendicular to another, the most usual as well as the most easy method is to make the joint square, with a short tenon of about one-fourth of the thickness of the framing to retain it in position.

Care must be taken to cut the joint accurately, or the pressure will bear solely on the projecting parts. It has been suggested that the end of the perpendicular post should be convex and fit into a concave sinking in the horizontal beam; although this would allow the joint more *play*, it has been proved by Hodgkinson that in long pillars which are flat and firmly bedded the resistance to fracture by flexure is three times greater than when rounded and capable of turning, though this is somewhat less in short pillars. When it is required to join two pieces of timber, not at right angles, the abutment if possible should be perpendicular to the strain. In the case of a principal rafter joining the tie-beam, the rafter should be cut into the latter about one-half its own depth, and the joint at the internal angle should be left a little open so as to allow for a slight settlement. The joint that the principal rafter makes with the king or queen post should, where possible, be square with the back of the rafter and should be provided with a short tenon; the joint at the external angle should be a little easy, to allow for settlement as before stated.

All dovetail joints should be studiously avoided in carpentry, and though they are not infrequently used in collar-beams of small roofs, &c., they are not satisfactory; and, as the slightest shrinkage destroys the strength of the combina-

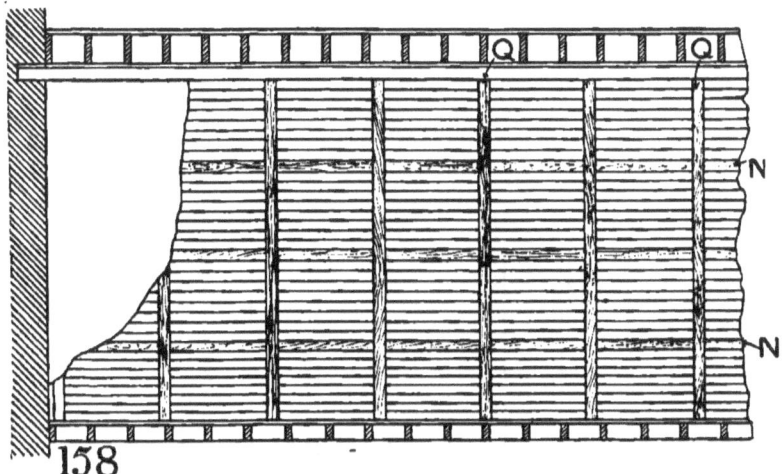

158. Bricknogged Partition.

tion, they are liable to crippling owing to the wedge-shaped cutting away at the shoulder.

PARTITIONS.—In carpentry, partitions refer to the frame timber-work separating different parts of a building from one another, and they are generally plastered or boarded. For convenience, they may be divided into—

1. *Brignogged partitions.*
2. *Quarter partitions.*
3. *Trussed partitions.*

1. BRICKNOGGED PARTITIONS consist of a row of vertical

posts or quarters (Q, fig. 158), generally about three feet apart, divided by horizontal nogging pieces (N, fig. 158), from 1 in. to 3 in. thick, which are generally fixed not more than 3 ft. apart. In common work and where space is a great consideration, the bricks are placed on edge, and thus the width of the partition is reduced to 3 in. Bricknogged partitions are generally used only on the ground floor, or where they have a continuous bearing, and of course when it is possible to build a 9-in. wall they should not be used. It is contended by some that they become damp when used

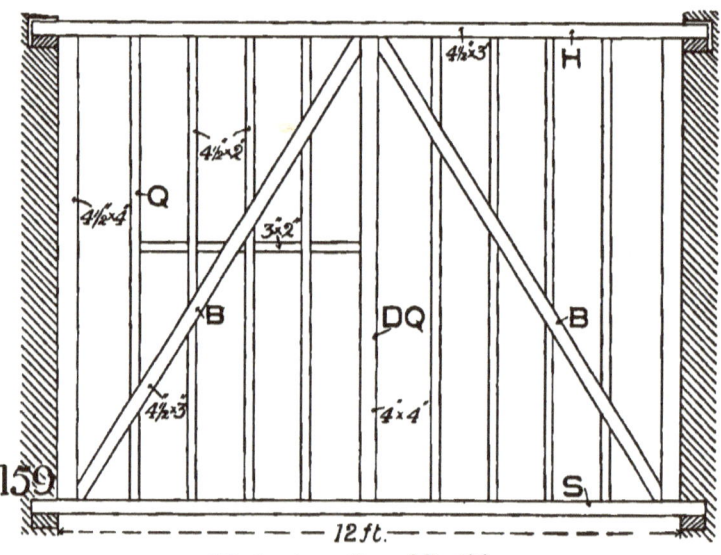

159. Quarter or Framed Partition

on the ground floor ; but if they are built on a good foundation with a proper damp course, this argument can hardly apply any more than it would to other walls. Bricknogged partitions do not require any braces, and are, in fact, better without them, as cracks from cross-strains are thus avoided.

2. QUARTER OR FRAMED PARTITIONS.—In cases where no doorway is required, these may consist simply of head H, sill S, and quarters Q, as shown in fig. 159, which may be stiffened by braces B, B, meeting against the head of a *double*

quartering DQ, and thus assisting to transmit the weight on to the walls. The quarterings are stiffened by nogging pieces every few feet. In fig. 159 the following scantlings would be used for a bearing not exceeding 20 ft. :—*Head*, 4½ in. × 3 in. ; *sill*, 4½ in. × 3 in. ; *quarterings or studs*, 4½ in. × 2 in. ; *double quarterings or principal posts*, 4½ in. × 3 in.; *braces*, 4½ in. × 3 in. ; *nogging pieces*, 3 in. × 2 in.

The ends of the head and sill should have a good bearing on fir wall-plates, and thus keep the weight off the floor, which probably would not be equal to carrying the additional load concentrated on to so small an area. The upper ends of the *quarters* should be secured to the head by iron

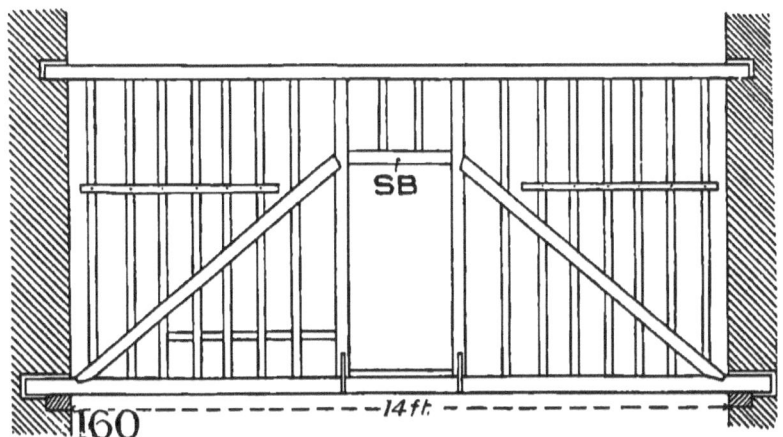

160. Quarter Partition, framed for one door.

straps after having been tenoned into it, and they should butt against and be nailed to the *braces*, and stub-tenoned into the sill. The braces should be housed into the sill, and also into the heads of the principal posts.

The studs should be about 1 ft. apart, centre to centre, which will allow of the laths being conveniently nailed to them.

Where a doorway is required in a quarter partition, the method generally adopted is that shown in fig. 160. In this case the braces are tenoned into the principal posts at the level of the straining beams, and it will be seen that by this method the weight is well transmitted to the walls. Tredgold advises that the braces should be placed at an angle of 40 deg. to the sill.

3. TRUSSED PARTITIONS are used in the place of quarter partitions where it is required to assist in supporting the floor above. Fig. 161 represents a trussed partition having a wide aperture in the centre to receive folding doors. It practically consists of two very strong trusses, the sill of the upper one being the inter-tie (I). This

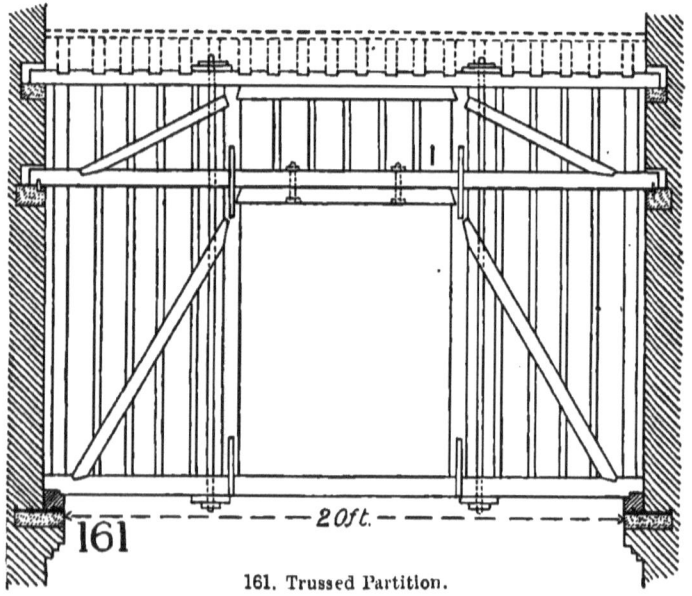

161. Trussed Partition.

is a very strong partition, well adapted to carry the floor above, and its strength is still more increased by the iron rods running from sill to head through the inter-tie, and securely bolted.

This partition is called *one-fourth trussed*, meaning that the upper truss occupies one-fourth of the whole height. One of the best methods of treating a partition where two doorways occur, one adjacent to each extremity, is shown in fig. 162.

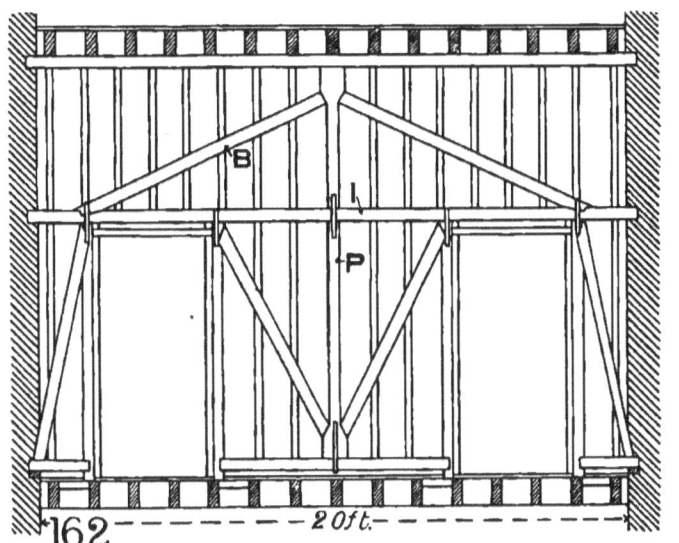

162. Partition framed for two doors.

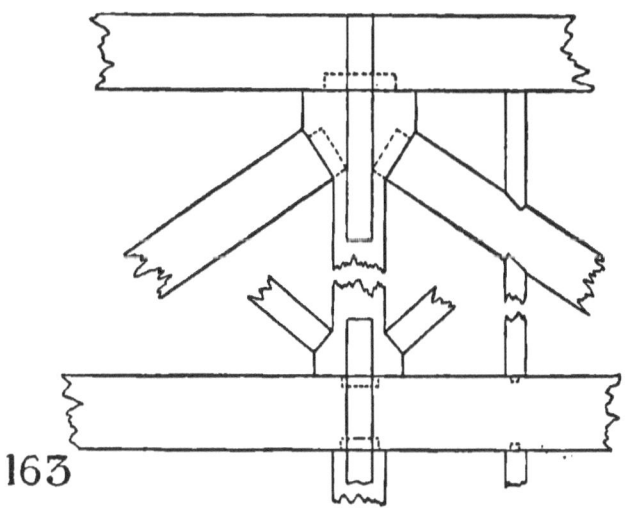

163. Detail of joints.

In this example the inter-tie I, the posts P, and the braces B form a kind of king-post truss. The door-posts are secured to the inter-tie by straps; the king-post is made *equivalent* to running through the two trusses, owing to the use of the strap shown in fig. 163. In this case it will be seen that the sill does not run through from wall to wall, owing to the floor-joists running transversely under it, and thus preventing it being placed between two of them. It will be seen that the whole system of framing is dependent upon the upper truss, which has not only to sustain itself and the lower portion, but has also to carry the floor above.

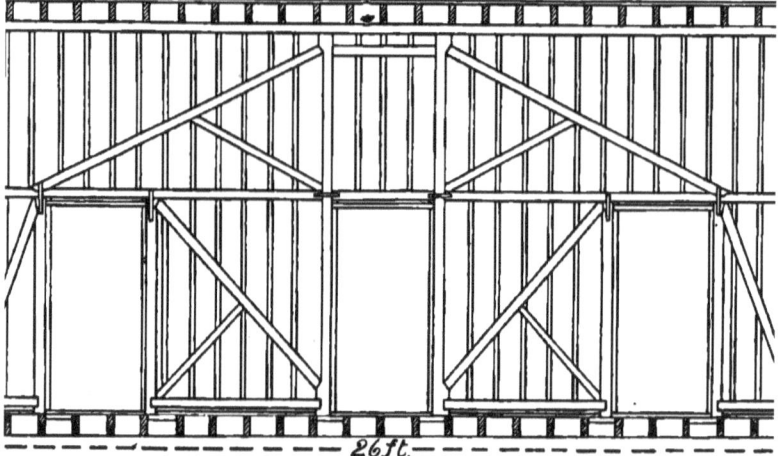

164. Partition framed for three doors.

The line diagram shown in fig. 164 represents a partition framed on the principle of the queen-post truss, and designed for the purpose of having three doorways, one in the centre and the other two at each end. The symmetrical arrangement of the doors renders it easy to place the pressure on the walls. The queen-posts run right through the inter-tie to the sill, and the inter-tie itself is connected by straps across the queen-post.

The braces, which correspond to the principal rafters in a queen-post truss, are stiffened by struts fixed against the inter-tie and queen-post, as shown. As in the last case, the stability of this partition is principally dependent upon the upper truss.

In cases where the door-posts do not come on to a joist, *firrings* are fixed between two of the joists upon which the extremities rest. Owing to imperfect joints, and the tendency of the timber to shrink, settlements frequently occur in partitions, and often serious cracks are found to occur after they have been plastered. It is very essential that only well-seasoned timber be employed, and the partitions should be left some time after they are erected, and with the whole ultimate weight upon them, before they are plastered, so that they may take their bearings, and any defect be made good. The arrises of all timbers exceeding 2½ in. in width should be bevelled off so as to admit of the plastering having a proper key. Tredgold gives the following data which will serve as a useful guide :—

	lbs. per square.
The weight of a square of partitioning may be taken at from	1,480 to 2,000
The weight of a square of single joist flooring, without counter-flooring	1,260 to 2,000
The weight of a square of framed flooring with counter-flooring	2,500 to 4,000

Scantlings for the *principal timbers* of a partition bearing *its own weight only*—

4 in. × 3 in. for bearing not exceeding 26 ft.
4 in. × 3½ in.   „   „   30 ft.
6 in. × 4 in.   „   „   40 ft.

If the partition has to sustain the weight of a floor or roof, the sizes of the timbers must be increased to meet the additional strain that will come upon them.

The filling-in pieces should be just thick enough to nail laths to, *i.e.*, about 2 in.

FRAMEWORK HOUSES are, as a rule, only erected in this country at the present time where other building materials are expensive, owing to the cost of carriage, &c., or in the case of half-timber houses, and sometimes for tile-hung exteriors. Wood-framed walls are also frequently used for such purposes as cricket pavilions, summer-houses, and such-like structures.

Fig. 165 shows the kind of building which is sometimes erected at the present day; the framing used in all such constructions being of a very similar nature.

The sound rule of construction of placing void over void and solid over solid, which by symmetry and correspondence of parts is always agreeable is as essential in timber as in any other kind of structure, and all the vertical timbers, even to the quarterings, should be so designed that no unnecessary strain be brought on the horizontal timbers.

In fig. 165, S shows the ground sill, P the principal posts, I the inter-ties secured by straps, all the vertical timbers being in the same plane. The ground sill receives the tenons of the principal posts, the posts in return receiving the tenons of the bressumers or inter-ties, which latter carry the floors. At the top the head or crowning beam is mortised to receive the tenons of the posts. The braces divide the parallelograms formed by the vertical and horizontal timbers into triangles, thus stiffening them, and helping to keep their original form. Between the bressumers the door or jamb posts are framed, as are also the window posts. The horizontal pieces framed into these, to form the heads of the openings, are called transomes or lintels, and there are also the sill pieces of a similar construction. All beams which are framed into the principal posts should be strengthened by an iron strap, and in the case of those returning to form the side-walls, there should be a right-angle strap going round the post.

In half-timbered work, the framing of the carcase being complete, the interstices are filled in with small stones set in mortar or concrete, the timbers having been previously dressed on their exposed faces.

In many half-timbered houses the ground-floor story is of brick or other fire-resisting materials, and the wood sill is placed on the wall at the first-floor level, or is projected out and supported on the joists of the first floor. Where half-timber is not adopted, the face of the framing, interior and exterior, is lathed and plastered, or faced with wooden

CARPENTRY AND JOINERY. 151

165. Framed House.

boards. The space between the inside and outside face is often filled up with some non-conducting material.

Instead of lathing and plastering, the exterior face is often covered with feather-edged boarding, as shown in fig. 165, and wooden cornices and other features are added.

In America timber-framed houses are rather the rule than the exception in the country districts. Being substantially built, and the spaces between the inner and outer faces specially treated with slag-wool and double boarding, they are preferred by many as being of a more equable temperature than ordinary built houses. In Sweden a dwarf wall, about 2 ft. high, of granite is generally built; on this the sill is placed and the posts, about 8 in. square, are mortised to it, and the rest of the construction proceeded with as previously described. The framing is then covered on the inside with $\frac{3}{4}$ in. deal boarding, and the outside with two thicknesses of the same, the first being laid horizontally, the second vertically. The joints of the latter are covered by fillets nailed along their length. The void between the two faces is filled with shavings or moss. More elaborately framed houses may be constructed on the principles enunciated previously under trussed partitions.

CHAPTER XV.

ORNAMENTAL CARPENTRY.

UP to the present our articles have dealt principally with carpentry as a constructive science, and with comparatively little regard to artistic merit. The reader, on referring back, will observe that constructive necessity has initiated and controlled the shapes of the various types, and that in many, if not most cases, the constructions were to have been built in an edifice, and when fixed were not to be visible. In such cases, it is but natural that little care should be expended on the appearance, and, in fact, it would be waste of time and labour to ornament it. Works on carpentry invariably leave out the decorative side of [the craft, and therefore it has been thought advisable to include in this book a chapter in which the setting out and construction of some of the most usual types of decorative carpenter's work are explained. We must guard the reader, however, against the idea that we are offering these examples, even the ancient ones, as models of design ; they are merely given as illustrations of the type of work which is referred to.

HALF-TIMBER WORK.—In the middle ages, houses formed of timber framing were common, and perhaps usual, in both town and country, and many are the picturesque groups of cottages and town fronts which are rapidly disappearing from our midst. In London the Staple Inn, Holborn, still remains as an admirable example of this type, and should be studied in its detail for the true principles of this type of building, of which it is in every way an admirable example.

The fire of London in 1666 must have been responsible for the loss of a large number of important timber-framed houses. In Chester are an admirable series of such house fronts, many of which have from time to time been illustrated in the *Builder*. Houses on a larger scale are also numerous, especially throughout the counties of Lancashire and Cheshire,

A small illustration of a gable to a house at Hereford is given as an example (fig. 166), showing the curved braces which are inserted as stiffening pieces, and whose design in many cases is elaborate and effective.

Oak should of course be employed, where possible, as solid framing, but where this is impossible, it is often used as a thin facing ($1\frac{1}{2}$ in.) to fir posts and rails behind. The general framing of these buildings is composed in modern work of upright posts and rails of timber, $4\frac{1}{2}$ in. by $4\frac{1}{2}$ in.,

166. Gable to a house at Hereford.

or 6 in. by 6 in., mortised and tenoned into each other, and pinned where necessary with oak pins. Curved braces are inserted not only for strengthening the framing, but also as an aid to the design, as shown in the illustration.

Oak has often been tarred in old houses for the sake of preservation, and presents a very quaint appearance as seen in Lancashire and Cheshire; or the surface may be left its natural face. If pine is used it may be treated similarly or painted. A quaint and picturesque effect is obtained by allowing the floor timbers of the upper story to project, as in the illustration; these are cantilevered over the main face of the framing beneath, and are likewise often aided by curved brackets which support a plate carrying their ends, the face of the wall above being carried up vertically from its outer plate, as in the illustration. The method of constructing the outer walls of these half-timber buildings has to be carefully considered; the spaces between should be filled in with concrete or brickwork, and have an inner face of slag-wool. A plan is given of a method (fig. 167) which has been used in modern buildings, and which has been found

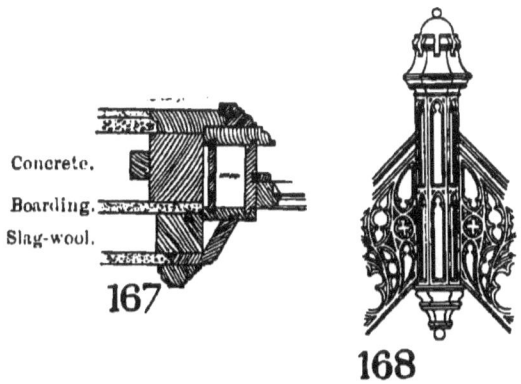

167. Plan of framing. 168. Finial from Coventry.

successful in keeping as equable a temperature as is possible in buildings of this kind. It will be observed that beside the concrete filling there is an inner compartment of slag-wool.

DORMER WINDOWS form features which aid in producing a picturesque skyline, and are, in many examples, mediæval and modern, very elaborately treated. Being part of the roof, they are almost invariably of timber construction, formed into small gables. They are useful for lighting rooms entirely or partly in the roof. The skeleton construction of one of these is shown in a previous chapter on "Roofs."

L 2

BARGE-BOARDS form a necessary termination to a timber roof and are usually employed not only for half-timber houses but also in other houses of a domestic type; though it must be admitted that those of the more "ornamental" class are too often, whether in ancient or modern work, little better in an architectural sense than what may be called "gimcrack." Still, there is a desire for them at times, and if made at all they should be well and carefully made. The purlins, ridge, and wall-plate project beyond the face of the gable from 1 to 2 ft., and support the rafters.

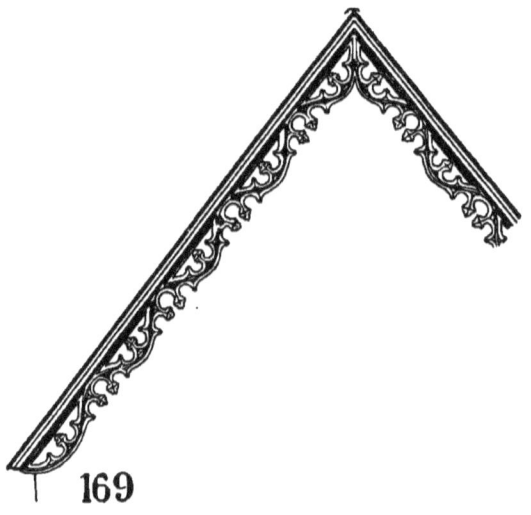

169. Barge-board from Droitwich.

The end rafter is covered with what is known as a barge or verge board, and the tiles or roof-covering project 1 in. to 2 in. beyond its face. Barge-boards, if over 11 in. wide, should be framed. In certain cases they are tenoned at the ridge into an upright post called a finial, and are often elaborately treated, as in the example from Coventry, fig. 168, in which a portion of the barge-board is also shown on each side. An open-work barge-board is also given, fig. 169, from an old house at Droitwich, and a better-designed example fig. 170, from a cottage at Worsborough, near Barnsley, in Yorkshire. It shows a rather elaborate specimen of a barge-board, carved beam, and gable finial. In the walls in old work, the face timbers occupy as much room as the intervening spaces, which gives a look of solidity, sometimes wanting in modern examples. This close spacing should be adopted; it is shown in the present illustration.

TURRETS and features of a like nature are often formed on the summits of roofs, both for use and ornament. They may contain an extract ventilator from one of the rooms below, in which case they are treated so as to form an

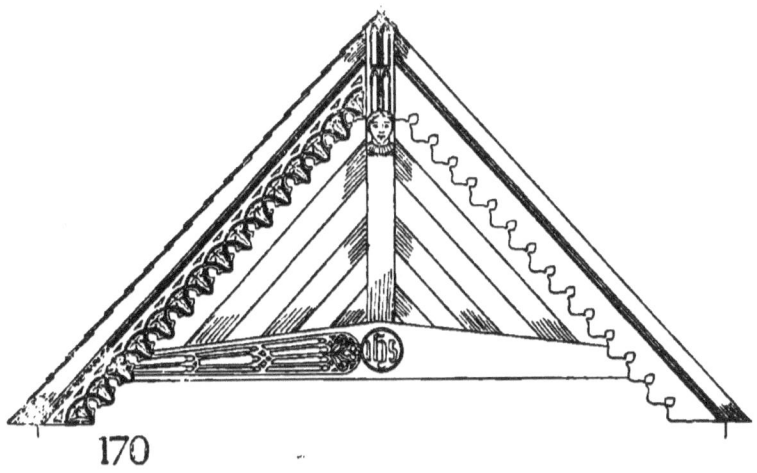

170. Gable-end from Worsborough.

artistic screen. A turret of this kind often contains a bell. A plan, elevation, and section of a design for one of these is given (figs. 177 and 178), a reference to which will sufficiently show the construction. As a feature of this kind is exposed, it is necessary to be braced sufficiently to enable it to withstand considerable wind-pressure, and it must, therefore, be cross-braced in addition. It will be noticed that the rafters to the upper part are framed into a central post, through which a tie-rod passes, receiving the iron weather-cock. The upper portion is sufficiently ventilated by means of the small louvres. The central post is held in position by two pieces of stuff passing on each side of it, holding it to two sides of the octagon. The framing to support the bell is indicated. The posts to the octagon are out of 4 in. stuff; that forming the arching is 2 in. thick, tongued into the framing; the shaped truss at the bottom is out of 2¼ in. stuff, grooved and tongued into the posts, and held firm by oak pins. The cornice is made up as shown.

DOVE-COTS are framed up in a similar way. Figs. 179 and 180 show an example lately erected over some stables, where it was necessary to arrange for a ventilating shaft in

171. Porch from Church of Huddington. 172. Plan of porch.

the centre. It will be seen that it is of very simple construction, being 4 ft. square over all. The angle posts are 5 in. by 5 in., intermediate posts 4 in. by 4 in., framed at top and bottom into rails pinned with oak pins. The angle posts are carried down and framed with strong rafters 6 in. by 4 in. The pigeon-holes are framed out of 1¼ in. boarding, and the ventilating shaft is also of 1¼ in. rough boarding, ploughed and tongued, and made air-tight with double-hinged and weighted flaps at bottom. The rafters are 4 in. by 3 in., covered by rough boarding, spaces at the top being

CARPENTRY AND JOINERY. 159

left for ventilation as indicated. The central post into which the rafters are framed is fitted between two horizontal struts, and the iron finial is held in position by a rod passing down the centre and bolted. The example is not in any way given as a model to be copied, but merely as indicating a feature in which craftsmen may expend a little more interest than in the rougher kinds of work.

PORCHES have frequently to be erected, not only for domestic but also for ecclesiastical work. They are specially suitable for the entrances to half-timbered houses. The example from the village church of Huddington, Worcester-

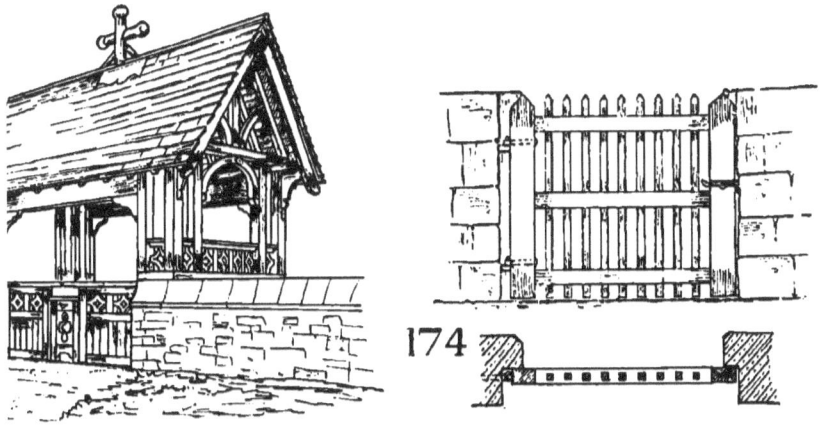

173. Lychgate.   174. Garden Wicket.

shire, is shown in fig. 171, and fig. 172 shows plan. The porch itself is 6 ft. 10 in. wide, 6 ft. 2 in. deep, and the front and back posts are out of stuff 11 in. by 6 in.

LYCHGATES are an especial type of work which lend themselves easily to the elaboration of the skilled carpenter. As a temporary resting-place for the coffin, any thing very ornamental might be considered out of place; at the same time a dignified, substantial, and not too plain treatment leaves plenty of scope for design. A shelter is indispensable to a lychgate, and is generally supported on curved braces tongued to the posts, and held in position by oak pins, which are allowed to project and tell their tale. The illustration (fig. 173) will explain the general principle of these constructions.

CONSERVATORIES.—Although these have often not received the attention they deserve there is some scope for the carpenter. An end elevation is given of one lately erected, in which an attempt was made to improve on the ordinary stock pattern (fig. 175). The illustration shows the posts, out of stuff 6 in. by $4\frac{1}{2}$ in., and the casements, and also the moulded rafters and sliding sashes.

175. End elevation of Conservatory.

GATES have to be very carefully put together in consequence of the comparatively rough wear they have to undergo by opening, shutting, and slamming. The jointing should be carefully performed, and the horizontal timbers should be as continuous as possible, as the strain is mostly on them. A detail of a gate (fig. 181) is given, forming an entrance to a drive. The method of firmly securing the posts is shown.

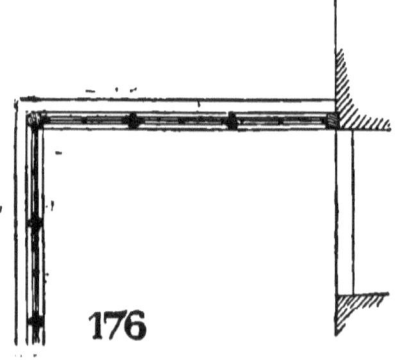

176. Part of plan of Conservatory.

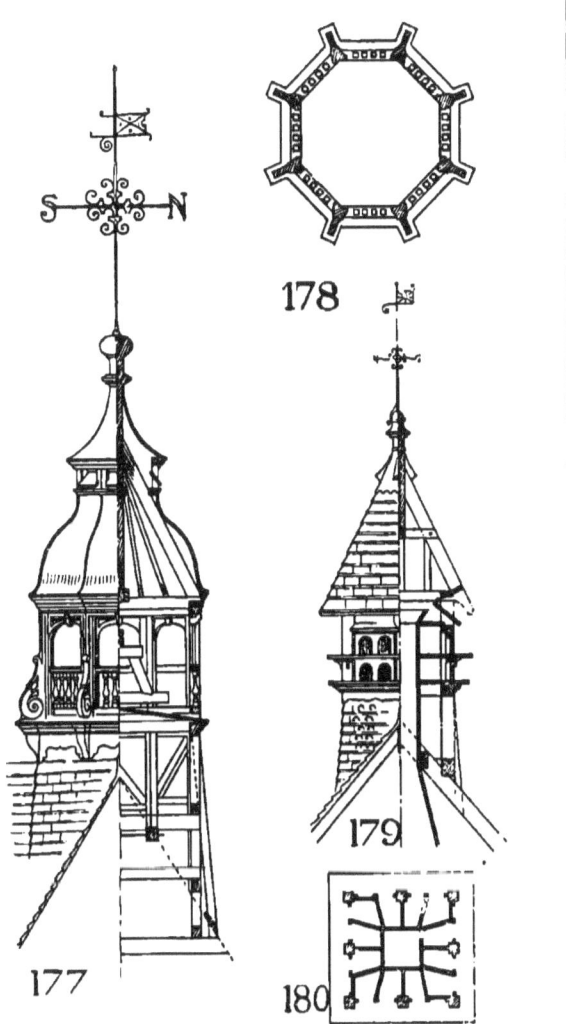

177. Turret, with bell. 179. Dove-cot.
178. Plan of turret. 180. Plan of dove-cot.

These are 9 in. square, and are taken below the ground to a depth of 4 ft. 6 in., and tenoned into a continuous sill, and held in position by struts, as shown. The gate itself is out of 4 in. oak, and is 3 in. thick for the

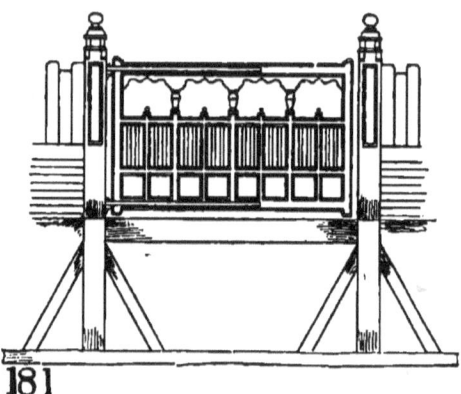

181. Entrance gate to a drive.

principal timbers, the arching and the smaller uprights being of $1\frac{1}{2}$ in. stuff, mortised and tenoned. The oak paling at the side is started to show the junction.

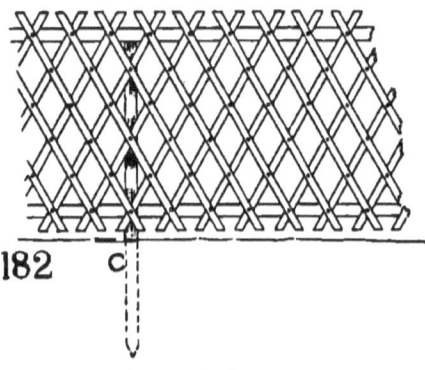

182. Lattice fencing.

A GARDEN WICKET is shown in fig. 174. The gate is hung to iron staples in the wall by means of iron bands with looped ends, fixed to the style.

PALINGS.—Several types of these come under the carpenter's practice. They may be divided into (a) lattice fencing, (b) lapped paling, (c) open paling. Lattice fencing, fig. 182, consists of a number of laths, a, pegged across each other, and supported by rails, carried on posts, c, fixed at intervals of about 10 ft.

Lapped paling of cleft oak is shown in fig. 183; the pales, A, lap over each other, and are nailed to rails, B, tenoned into posts, C, a skirting board, D, being fixed along the bottom to serve as a finish.

Open paling is shown in fig. 184. Posts, C, are planted firmly in the ground from 8 ft. to 10 ft. apart; rails, B, two or more in depth, are tenoned into these, and on them the pales are nailed flat.

SUMMER-HOUSES are objects on which the student may

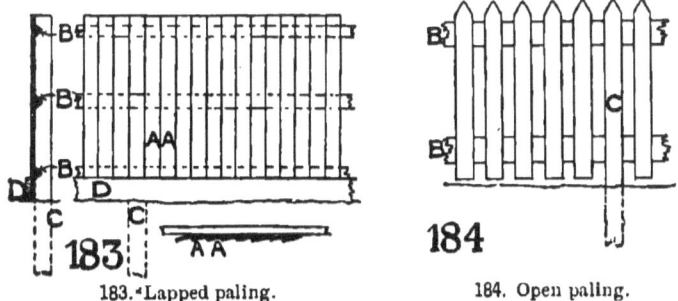

183. Lapped paling.   184. Open paling.

lavish a considerable amount of thought. Being of small dimensions, simplicity should be regarded as essential. The attempt so often made to *rusticate* these adjuncts to a garden may be regarded as pedantry. On the other hand, natural timbers may well be used. The most suitable are oak, fir, and larch. In landscape gardening, a well and simply designed summer-house of unsawn wood may be employed in out-of-the-way corners, but not made ridiculous by affectation, as if it belonged to some half-civilised person. As to the construction of these, if the reader has carefully followed these articles he will have grasped the methods of joining timber suited to any emergency. Such constructions may be suitably roofed with oak shingles, as being in harmony with the surroundings; these should be torn or rent from the log and not sawn, in order to stand the weather.

CARPENTERS' FURNITURE.—Although this may be deemed hardly to come within the scope of these articles, yet seeing that there has been a tendency of late years to drag the designing and execution of furniture out of the "quagmire" into which Tottenham Court Road has drawn it, and to take it, as it were, out of the upholsterers' hands, it is considered that it might fitly end this chapter.

Of all things which the architect may design, furniture will assuredly teach him the most. The sizes are practically fixed, and he must keep to them, which as a training in regard to fitness is excellent. Of benches, tables, chairs suitable for halls, and having more of the stability and solidity than is usual nowadays, there are many examples of the Elizabethan and Jacobean period in the South Kensington Museum and elsewhere which should be measured and drawn to scale by the student; fig. 185 is an example of the true principles of construction, although the decorative portion of the design cannot be recommended in every respect as an example. Individuality should be stamped on furniture as on the carcass of the house itself; and in reality there seems a tendency, which we hail with satisfaction, among the more educated classes, to allow the architect to design the whole of the internal fittings of his house. In the past, such fittings were a necessary complement of the architecture proper. What would the interior of an Elizabethan Hall be without its carved overmantel, its characteristic doors and enrichments, its sturdy table—not covered with a cloth to hide its imperfections—and last, the richly-moulded arm-chairs made to last, not for the date of a passing fancy or fashion, but for the life of the structure itself.

185. Chair, French, fourteenth century.

CHAPTER XVI.

JOINTS USED IN JOINERY.—HINGING.

IN treating of this section of our subject the name is an evidence of its importance, for joinery is the art of joining and framing wood for the internal and external finishing of buildings. Newlands distinguishes carpentry from joinery by this, that while the work of the carpenter cannot be removed without affecting the stability of a structure, the work of a joiner may. Being of a finer description a much greater care and precision is required in joinery than in carpentry. Joinery then, as we understand it, is comparatively a modern art, the first evidences appearing in the pulpits, screens, thrones, and stalls of our Gothic cathedrals. In the earlier samples the work is of such a simple nature that it is surmised that the two crafts were then one, and that only in later times has the elaboration of fittings tended to separate them.

The first work on the subject was in 1677, quaintly entitled, "Mechanic Exercises on the Doctrine of Handy works."

To proceed at once, however, to the consideration of joints in joinery, it may be mentioned that several of the most common of these have already been referred to under Chapter XIII. Floor-coverings. Following the system of classification we have adopted throughout, joints may be roughly classed under—

I. Joints between boards in the same plane.
II. Joints between boards meeting at a right angle.
III. Joints between boards meeting at other than a right angle.
IV. Joints between boards in which one is at an inclination to the horizon.
V. Joints for various purposes.

I. JOINTS BETWEEN BOARDS IN THE SAME PLANE.—By referring to Chapter XIII., such forms as the following are discussed: Plain or butt joint; rebated; rebated, grooved, and tongued; ploughed and tongued; grooved and tongued; rebated and filleted; and dowelled, all of which are used in the laying of floors, and may be equally used in other positions for connecting boards in the same plane.

Match-boarding (fig. 186) consists of boards placed side by side, and fitted into each other by means of a groove formed in one board, into which a tongue formed on the adjoining board fits. The "quirked bead" is stuck on the angle of one board so that in shrinking the joint may not be unsightly, but forms a symmetrically shaped moulding.

V-jointed boarding (fig. 187) has the edge of each board chamfered and grooved, and a loose wooden tongue is inserted to keep each in position, the junction of the two chamfered edges forming a V on plane, hence the name.

Slip-feathers are thin pieces of hard wood, such as oak, inserted into the length of the boards in grooves cut to receive them. For strength they should be cut across the grain, as otherwise they are liable to split along their length.

Mortise-and-tenon joints (fig. 188) are used in joiners' work similarly as in carpenters', the only difference being that they have to answer somewhat altered circumstances, and to be executed with the greatest regard to accuracy and finish. The joint is used in doors, and framing generally, and the tenon should be made about one-third of the thickness of the stuff on which it is formed; the width should not be more than five times its thickness, as, if more, it will be liable to bend. The dotted line shows a method of what is called "haunching" a tenon by leaving a projecting piece $\frac{1}{2}$ in. to 1 in. in length, which gives the tenon greater strength to resist any side shock.

Double tenons (fig. 189).—This form of tenon, which the drawing sufficiently explains, is used in wide pieces of framing in order to weaken the style in which the mortise is framed as little as possible. Two tenons of comparatively small width do not shrink so much as one large one. This method is also used for the lock-rail of doors, and is provided with a haunch between the tenons, which strengthen them.

II. JOINTS BETWEEN BOARDS MEETING AT A RIGHT ANGLE.—*a. Mitre-joints.*—The most common form of

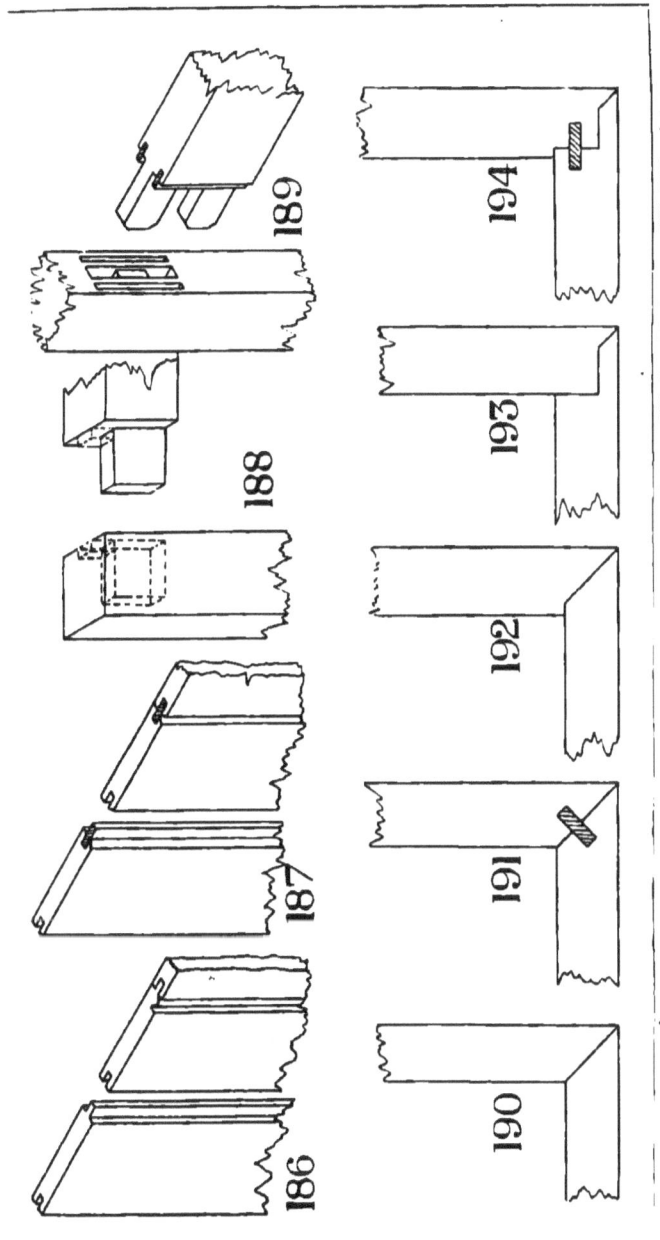

186. Match-boarding. 187. V-jointed boarding. 188. Mortise-and-tenon joint. 189. Double tenon. 190. Mitre-joint. 191. Mitre-joint with slip-feather. 192. Joint between boards of different thicknesses. 193. Rebate and mitre-joint. 194. Rebated joint with key.

uniting two boards at an angle is shown in fig. 190, known as the mitre-joint. Each edge is planed to an angle of 45 deg., and glued to the other. In order to make a stronger joint, a slip-feather is introduced, as in fig. 191. In the case of boards of different thickness, which are often used in conjunction, the joint is effected as shown in fig. 192, in which the mitre on the thicker stuff is formed on to the corresponding depth on the thinner stuff, the joint therefore being a combination of the mitre and simple butt-joint. Fig. 193 shows another method in which one of the boards is rebated, and a small portion of each is mitred; this joint may be strengthened by screws each way.

In fig. 194 both boards are rebated and a slip-feather inserted as a key. Figs. 195 and 196 are combinations of grooving and tonguing with the last-mentioned joint; they are, however, seldom used. In all these joints, the boards meeting at an angle, the slight opening at the mitre caused by shrinkage would be scarcely noticeable.

When, however, two pieces of stuff butt against one another, a *butt-joint* being formed, it is evident that the shrinkage would cause an open joint. For this reason a bead-moulding is run on the edge of one of the pieces as in fig. 197, where one board is rebated from the back and a bead is formed on the external angle of the abutting board.

Fig. 198 shows a joint in which a groove is formed on the inner side of one board, into which a tongue on the edge of the other fits. Fig. 199 is similar to the last except that a bead and cavetto is run on the internal angle of one board and a cavetto on the external angle of the abutting board, this would somewhat cover the shrinkage of the former board.

Fig. 200 shows a "quirked bead" run on the external angle of one board and the abutting board is simply rebated as shown. Figs. 201 and 202 show joints used in the formation of cisterns, in which, for strength, one board is continued along the face of the adjoining board.

b. Dovetail-joints.—In better class work this form is used in preference to a plain mitre-joint. There are three kinds of dovetail : the common, the lapped, and the lapped and mitred,

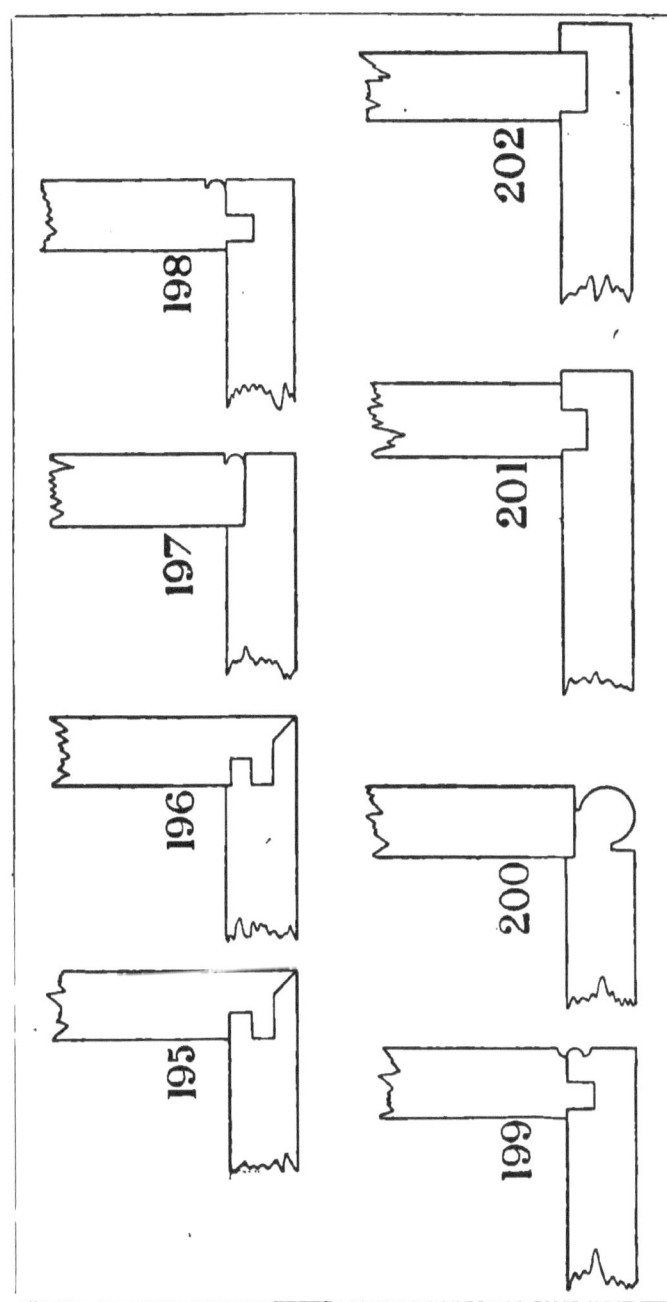

195 and 196. Grooving and tonguing. 197. Beaded butt-joint. 198. Beaded butt-joint with tongue. 199. Butt-joint with cavetto moulding. 200. Butt-joint with quirked bead. 201 and 202. Joints used in the formation of cisterns.

The common dovetail is shown in fig. 203. It will be seen that the junction is formed by alternate projections and indentations of a dovetail shape formed on one piece of stuff which fit into corresponding spaces on the other. In the figure the pins and sockets respectively are marked **P** and **S**, and are seen alternately on each side of the angle. The strongest joint is when the pins and sockets are equal in width, but for cheapness the pins are often placed much farther apart. Such a form of joint is used for the back angles of drawers, which do not show on the face. In hard wood, such as oak, the angle of the dovetails may evidently have more splay than in soft wood, such as deal, where they would have a tendency to split off on any strain being applied.

The *lapped dovetail* is shown in fig. 204; in this the dovetails are seen only on one side, the front face not being interfered with, as this portion is not cut through by the dovetails. The joint is generally used for drawers or other pieces of joinery in which the front is desired to be kept plain.

The *lapped and mitred dovetail* is shown in fig. 205; it is used in cases where it is intended that the dovetails should not be visible on either side or on the top of the joint. About two-thirds of the joint is usually dovetailed, the rest being mitred as shown.

III. JOINTS BETWEEN BOARDS MEETING AT OTHER THAN A RIGHT ANGLE.—Figs. 206—209 show methods which are adapted to boards meeting at an obtuse angle, which will sufficiently explain themselves. Such joints are useful in many positions.

IV. JOINTS BETWEEN BOARDS IN WHICH ONE IS AT AN INCLINATION TO THE HORIZON.—In certain cases, as in hoppers, tubs, &c., such a joint as indicated above is necessary. It is of a dovetail character, and is performed as indicated in fig. 210.

V. JOINTS FOR VARIOUS PURPOSES.—It is evident that joints for various purposes, not mentioned above, have to be designed as occasion arises. In window backs, wall-linings, &c., where several widths of boarding are united together by ploughed and tongued joints, a piece of wood, called a "key" (fig. 211), is inserted across the grain in a dovetailed

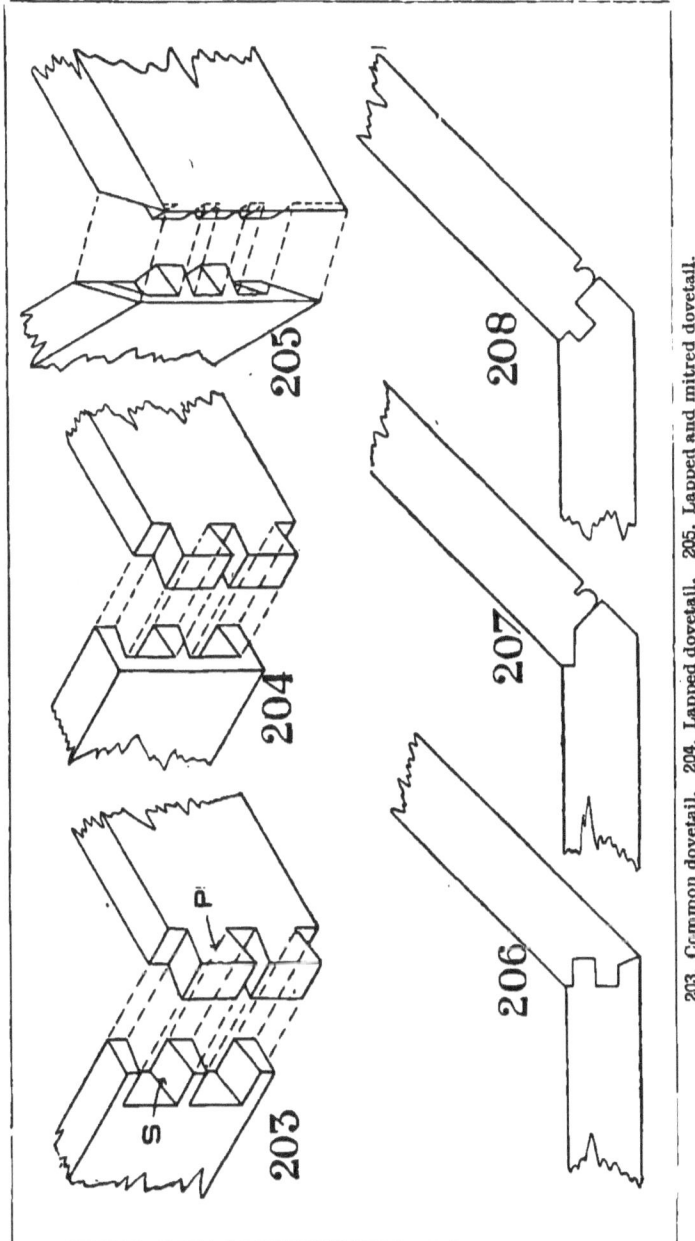

203. Common dovetail. 204. Lapped dovetail. 205. Lapped and mitred dovetail. 206-208. Joints between boards meeting at other than a right angle.

groove prepared to receive it, and prevents any tendency the boards may have to get out of their proper plane.

Clamping, both plain and mitre, is shown in figs. 212 and 213, in drawing-boards and such like cases, boards are kept in position by being framed into a clamp C, placed at the ends across their grain, and grooved to receive a tongue left on their ends, part being carried through as a tenon. Mitre clamping (fig. 213) is simply performed for the sake of appearance.

Glued Joints are performed by butting two pieces of wood together, their edges being previously well cleaned and dried and covered with a thin coating of glue and rubbed together. If properly carried out, such a joint, with no other aid but glue, is stronger than the wood itself, which, on pressure being applied, will crack anywhere but on the joint. Dovetail keys are placed across the grain at the back of such boards.

A *Glued and Blocked Joint* is used when two boards meet at an angle, as shown in fig. 214, and when, in addition to being glued, a block, B, is inserted in the angle. This considerably strengthens the joint, which is much used in the construction of stairs, as will be explained under that section.

HINGING.

The hinging or hanging of doors, windows, shutters, and so forth to their frames can now be touched upon.

There are many types of hinges and many ways of fixing each type, in order to make them answer certain requirements; but in all considerable care and judgment has to be exercised for proper adjustment.

For the sake of clearness they may be divided as follow:—

Butts are used for ordinary doors and windows which have not to clear a projection. Fig. 215 shows the hinging of a door to open at a right angle. Butt hinges are made of wrought and cast iron and brass, the forming varying in size from $1\frac{1}{4}$ in. to 4 in. in length, the latter from 1 in. to 4 in. The size used in practice varies with the weight and importance of the door to be hung.

A *Centre-pin Hinge* is shown in fig. 216, permitting the door to open either way, and fold back against the wall in

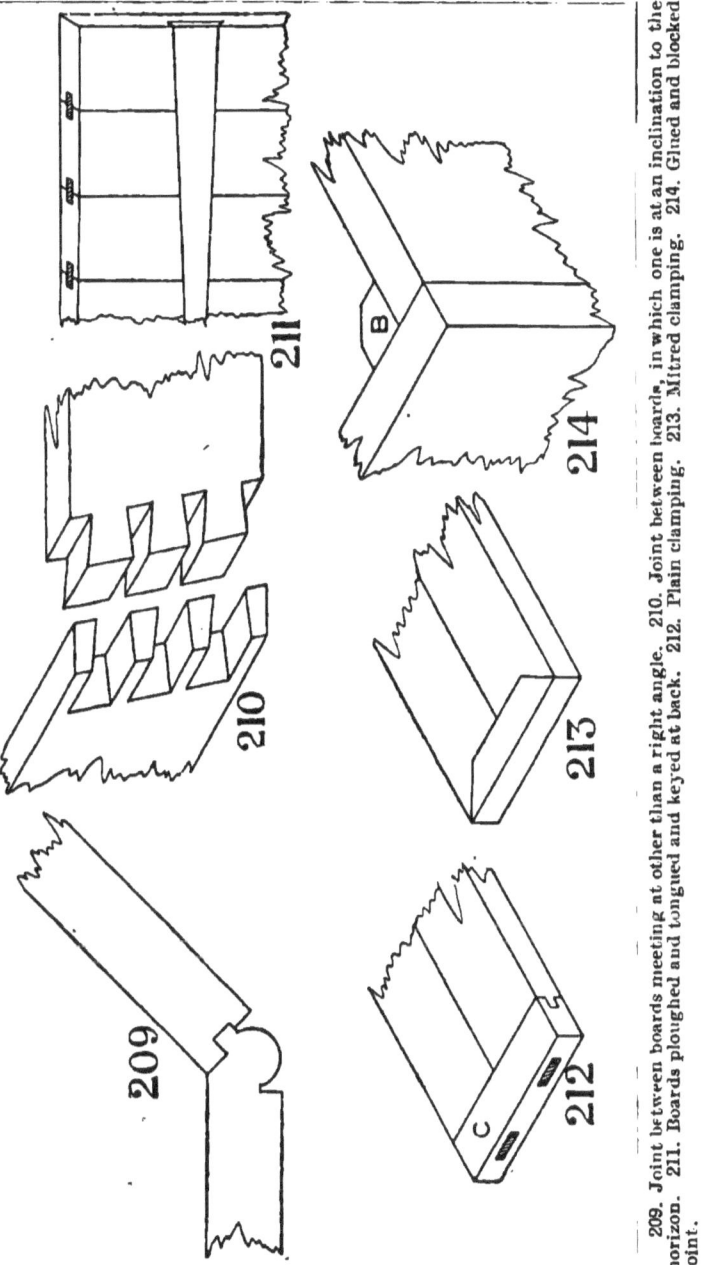

209. Joint between boards meeting at other than a right angle. 210. Joint between boards, in which one is at an inclination to the horizon. 211. Boards ploughed and tongued and keyed at back. 212. Plain clamping. 213. Mitred clamping. 214. Glued and blocked joint.

either direction. There are various other forms of centre-pin hinges, which need not be referred to.

Projecting Butts are used when a door, window, or shutter has to swing clear of a projection. Figs. 217 and 218 show a method of hinging employed when the flap, on being opened has to be at a distance from the style, as, for example, in shutters in reveal, or pew doors, and the like, which have to be swung clear of the mouldings of the capping.

Rising Butts, which make the door self-closing and also cause it to rise clear of a carpet, consist in the hinge having a portion of a spiral thread worked on to it, which, on opening the door, causes it to rise as stated above. The weight of the door causes it to work down on this spiral thread and shut itself by gravitation.

The Back-flap or Shutter Hinge allows of the leaves folding back against each other. Fig. 219 shows the hinging of a back-flap when the centre of the hinge is in the middle of the joint. Fig. 220 shows the manner of hinging a back-flap when it is necessary to throw the leaf back from the joint.

H and H hinges are sufficiently explained by their name; they are used for common work and for such cases as ledged doors or where stuff is too thin to screw butts on the edge.

Cross Garnets are in shape like the letter ⊢ placed on its side; they are used in the commonest form of external doors and are generally about 12 in. in length.

Hook-and-Eye hinges are used for gates and heavy outside doors, the eye being formed on the portion which is secured to the door and which fits over the hook which is secured to the frame.

A Rule Joint such as is required for a shutter is shown in figs. 221 and 222. These can be used when the piece to be hung is not required to open to more than a right angle. In such a case the centre of the hinge is necessarily placed at the point of juncture of the two pieces.

The division of our subject does not permit of describing or illustrating the elaborate specimens of wrought-iron hinges which form part of the door furniture of the Mediæval period, and which add so largely to the interest of the work of that period.

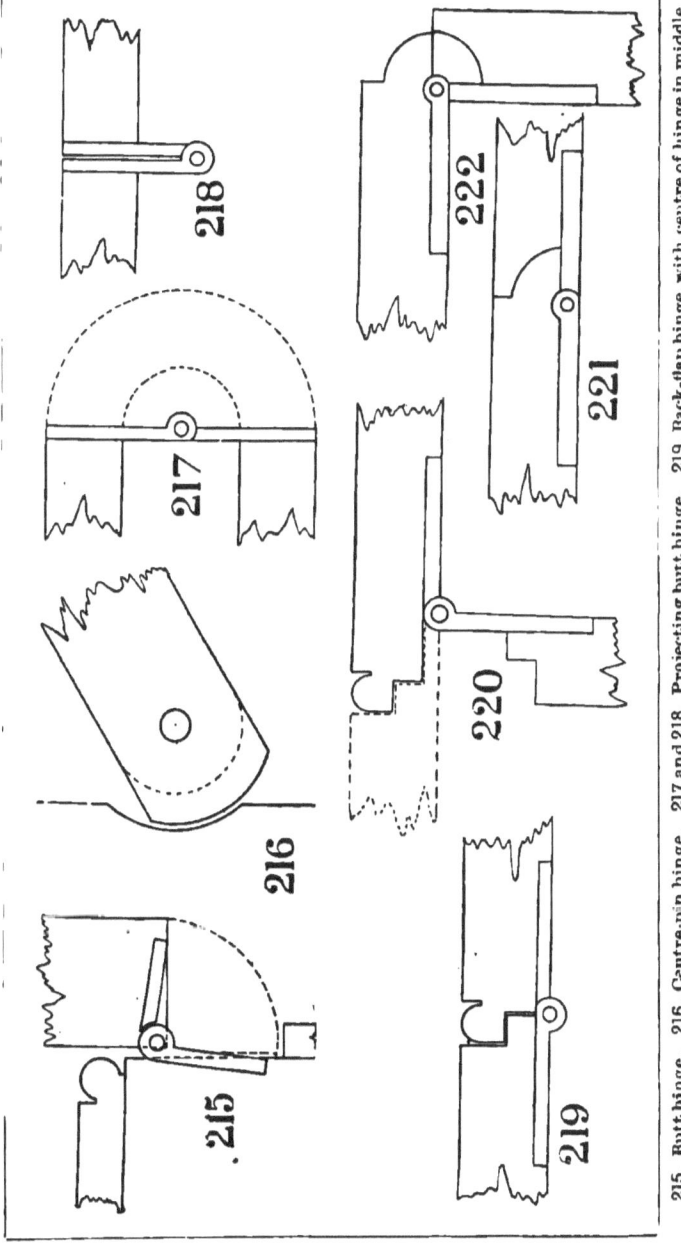

215. Butt hinge. 216. Centre-pin hinge. 217 and 218. Projecting butt hinge. 219. Back-flap hinge, with centre of hinge in middle of the joint. 220. Back-flap hinge, with centre of hinge back from the joint. 221 and 222. Rule-joint hinge.

CHAPTER XVII.

MOULDINGS.

MOULDINGS have been defined as the varieties of outline or contour given to the various members in a building.

In regard to carpentry and joinery we may almost say that nearly all the mouldings applicable to architecture generally are to be found in use, so that it becomes necessary to briefly name and illustrate these and the manner in which they are set out.

Any architectural member is said to be moulded when its edge presents continuous lines of alternate projections and recesses.

The subject may, historically, be classified as follows:—

I. Classic (so called) mouldings.
II. Gothic mouldings.
III. Modern mouldings.

I. CLASSIC MOULDINGS.—These are so called because they are derived from examples left us by the Greeks and Romans. The *fillet* (fig. 223) is perhaps the simplest form of moulding; in fact, it is so simple that it can hardly be said to be a moulding proper, but rather a small plain face to separate other mouldings. It is used largely in connexion with more elaborate mouldings, as in dividing groups from each other.

The *bead* or *astragal* (fig. 224) may be either a plain round formed on the edge of a piece of stuff as in the illustration, or, when formed so that the surface of the cylindrical part is flush both with the face and edge of the wood, a sinking being made on the face only, the combination is called a "quirked bead" (fig. 225). If a quirk is formed on each side of a bead stuck on the angle of a piece of stuff, as in fig. 226, the moulding is called a *bead and double quirk*. These form the simplest kind of mouldings, and are used in almost every conceivable position in carpentry and joinery, principally for the covering of the joint formed by two connecting pieces of stuff. A moulding called a *double bead*

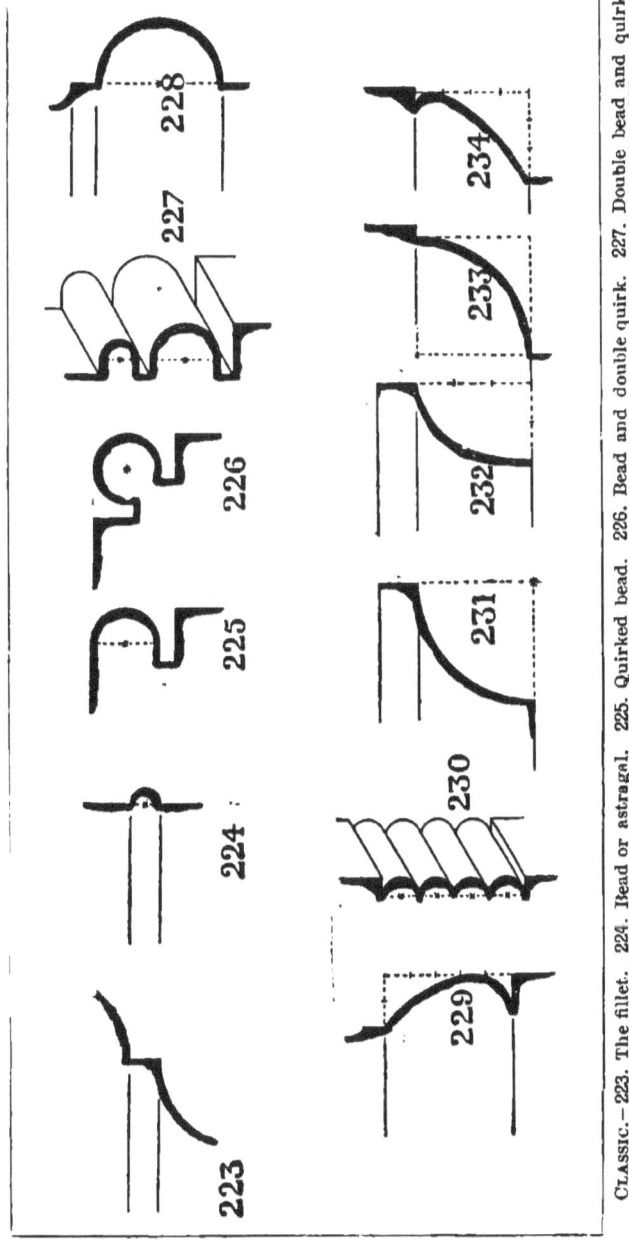

CLASSIC.—223. The fillet. 224. Bead or astragal. 225. Quirked bead. 226. Bead and double quirk. 227. Double bead and quirk. 228. Torus—Roman. 229. Torus—Greek. 230. Reeding. 231. Cavetto or hollow—Roman. 232. Cavetto—Greek. 233. Ovolo—Roman. 234. Ovolo—Greek.

and quirk is an extension of the above. It is shown in fig. 227, and consists of two semi-cylindrical mouldings, not necessarily of the same size.

The *torus*, fig. 228, is in many respects similar to a bead, the distinction being that it is always used in connexion with a fillet, whereas a bead is not so used. Fig. 229 shows a Grecian form.

Reeding, fig. 230, is so called when a succession of semi-circular mouldings are stuck on a piece of stuff.

The *cavetto* or *hollow*, fig. 231, is a quadrant of a circle, and is much used. Fig. 232 shows the more refined outline which is found in Grecian work.

The *ovolo* or *quarter round*, fig. 233, is described in a similar way, but with the convex side outwards. It is perhaps the most common joint used in joinery for sashes, doors, &c. Fig. 234 shows the same moulding as would probably be found in Grecian work possessing considerably more character, and being less mechanical in appearance.

All the above mouldings are comparatively simple, the Roman mouldings being struck from one centre, and forming distinct and simple curves.

We have now to consider mouldings having compound curves. Of these, the most frequent is the *cyma-recta* or *ogee*, fig. 235. In Roman work this was generally a curve of double curvature, formed of two equal quadrants and struck as shown in fig. 235. In Grecian work the moulding partakes more of the character of that shown in fig. 236, being of an elliptic form.

The *cyma-reversa* or *ogee-reversa* (fig. 237) is, as its name implies, simply the last moulding reversed or turned round. Such a moulding is largely used in cornices.

The *Scotia*, fig. 238, is a moulding originally used in Classic work, in the bases of columns in particular. The figure sufficiently explains its form, being composed of two unequal circular arcs, of which there are various methods of finding the centres. Fig. 239 shows the Grecian form, which takes an elliptical or parabolic curve, as in all Greek mouldings.

The *echinus* (or more properly the *ovolo*) of the Grecian Doric is set out in fig. 240. It is merely put in this form to

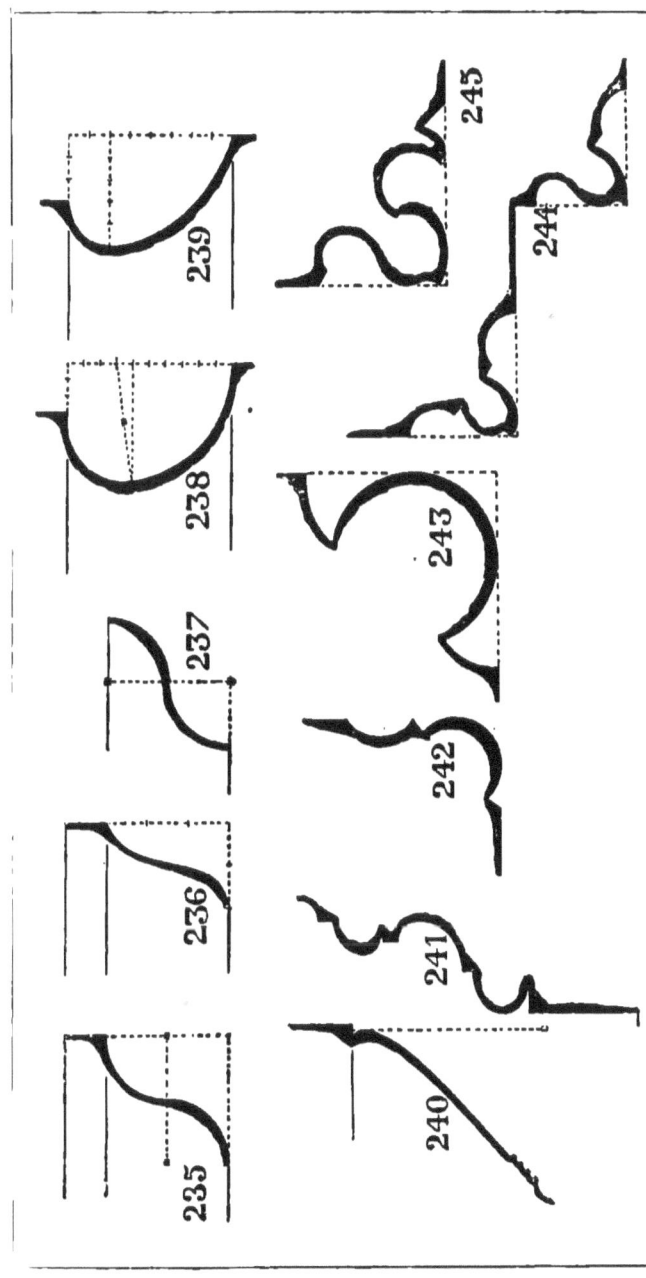

CLASSIC (*continued*).—235. Cyma-recta—Roman. 236. Cyma-recta—Greek. 237. Cyma-reversa or ogee-reversa. 238. Scotia—Roman 239. Scotia—Greek. 240. Ovolo or echinus—Greek. 241. Base mouldings. 243. Round bowtell or edge roll. 244. Keel moulding. EARLY ENGLISH.—242. Cylindrical edge roll and shallow hollow. 245. Pointed bowtell.

give the student its general outline. Fig. 241 shows the mouldings of a Greek base. The question of the drawing out of Classic mouldings has frequently been discussed. In Grecian work these mouldings approach very closely to conic sections—either hyperbolic, parabolic, or elliptic—but although this is so, it is not generally considered that they were set out mathematically, but merely drawn by hand, and that the artistic perception of the Greeks caused them to take shapes approaching these exact sections. Whether this is so or not, the result is evidently to be preferred to the mechanical hardness characteristic of Roman mouldings which were practically set out with the compasses and form parts of a circle.

II. GOTHIC MOULDINGS.—In Gothic mouldings applied to woodwork we find the same principles involved as in those executed in stonework, and, in fact, we can trace the period and style by means of the mouldings, similarly to the more substantial work in stone. In the Mediæval period, mouldings were executed on the solid framework forming the structure, and it was left to the later periods of the craft for mouldings to be run themselves and planted on afterwards. In Gothic structures, moulding was strictly confined as such to its legitimate sphere as an ornamental finish to the edges of the timber itself. For this reason it is interesting to study all old joinery, as the construction is apparent. For the Mediæval nomenclature of mouldings, the useful work published many years ago by Professor Willis, and followed by all writers on architecture, is adopted here. We can, therefore, divide the mouldings of the Mediæval period in the same manner as the periods of architecture of which they form such a conspicuous part.

In the *Norman period* the plain cylindrical edge roll and the shallow hollow, fig. 242, was the principal moulding, but few examples executed in woodwork have come down to us.

In the *Early English period* we have—
1. The edge roll or round bowtell.
2. The pointed bowtell.
3. The roll and fillet.

The *round bowtell* or edge roll is shown in fig. 243. The

earliest and most simple method of moulding was no doubt that of chamfering the edge; the next probably consisted in rounding the edge, which latter was probably developed into the round bowtell by cutting out a small angular channel on each side as shown.

The *pointed bowtell* is shown in fig. 245, and is generally taken as coeval with the introduction of the pointed arch. Its formation probably arose from a wish to emphasise the angles of recessed arches without interfering with the square edge. Another adaptation of this form is shown in fig. 244, in which it will be seen that a slight sinking is made close

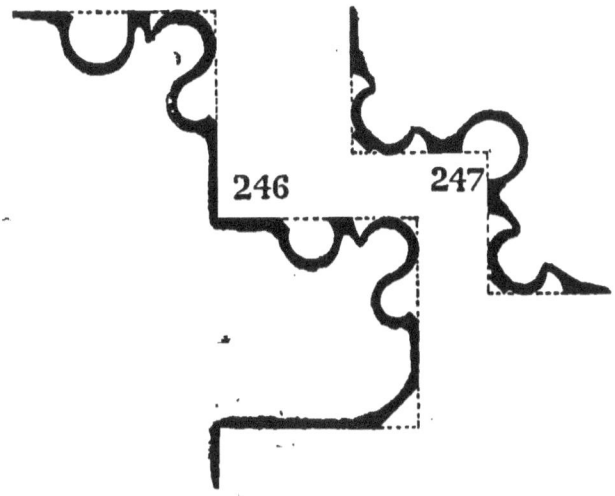

EARLY ENGLISH.—246. Roll and side fillet.
DECORATED.—247. Roll and fillet.

to the edge, so as to make the latter appear sharper and more distinctive. This form is also called a *keel moulding*, form the outline resembling the keel of a boat.

The *roll and fillet* is shown in fig. 254, and consists simply of the round bowtell and a fillet placed on the front, or it may be considered a derivation from the pointed bowtell with a fillet left on the edge instead of an arris.

A *roll and side fillet* is shown in fig. 246, which is a further

182   CARPENTRY AND JOINERY.

adaptation. The depth of the mouldings in this period constitutes the most characteristic difference between it and the succeeding style.

Decorated mouldings may be said to include, besides derivatives from the above, the following :—

1. Roll and triple fillet.
2. Ogee.
3. Double ogee.
4. Scroll moulding.
5. Wave moulding.
6. Plain or hollow chamfer.
7. Sunken chamfer.

The *roll and fillet*, fig. 247, is formed in this period with less undercutting than in the previous style, as is seen in

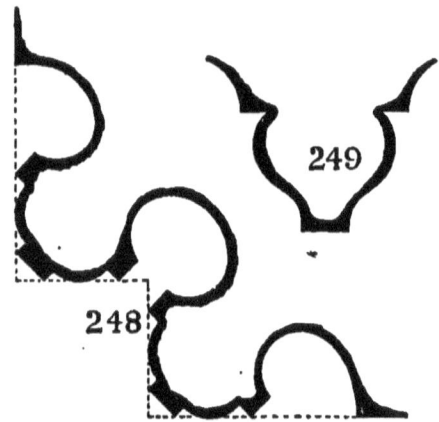

DECORATED.—248. Roll and triple fillet. 249. Ogee.

the illustration, and is generally accompanied by a hollow. The fillet itself is broader than in the Early English period.

The *roll and triple fillet*, fig. 248, is a development of the last named, and the usual form is explained in the figure.

The *ogee*, fig. 249, is generally considered to be a development of the roll and fillet, and, it will be seen, has not the same character as the classic ogee mentioned earlier. It will be noticed that the concave portion is not so large as the convex, hence the surmise of its development by rounding the edge formed between the roll and fillet.

The *double ogee* (fig. 258), divided by hollows, consists of two rolls and fillets, conjoined at their bases; it is one of the most usual mouldings in this and the next period.

The *scroll moulding* (figs. 251, 252), although in use in the latter part of the Early English period, is more distinctively

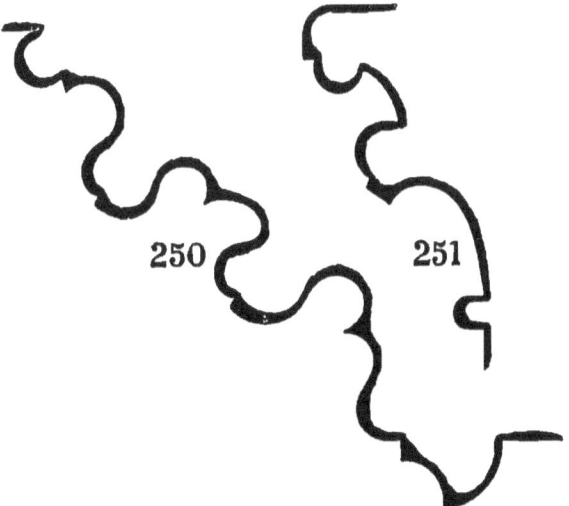

DECORATED. 250. Arch moulding. 251. Scroll moulding.

a Decorated moulding. Its name is derived from its resemblance to a roll of thick paper, the outer edge of which overlaps upon the side exposed to view. Another definition is "a cylinder, the under half of which is withdrawn or shifted a little behind the upper." The scroll moulding is practically universal in the abacus and neck of Decorated capitals, as in the figures mentioned above, and is frequently used in arch mouldings, as in fig. 250.

The *wave moulding* (fig. 253) is composed of two ogee curvatures, forming a central bulge or entasis. It is specially characteristic of the Flowing or Decorated period. Another variety is shown (fig. 261) in which part of the chamfer plane is left on. This, however, is most common in the next period.

The *plain* or *hollow chamfer*, figs. 255, 256, of two or more

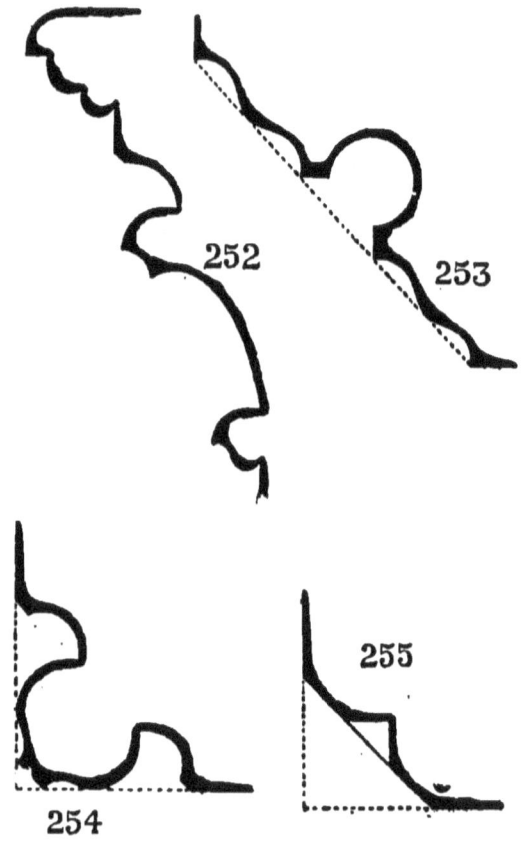

DECORATED.—252. Scroll moulding. 253. Wave moulding.
254. Roll and fillet. EARLY ENGLISH.—255. Plain chamfer.

orders, is frequently met with, being divided by $\frac{3}{4}$ hollows as shown in fig. 262.

The *sunk chamfer*, fig. 257, consists of a flat surface sunk between two raised edges on the chamfer plane.

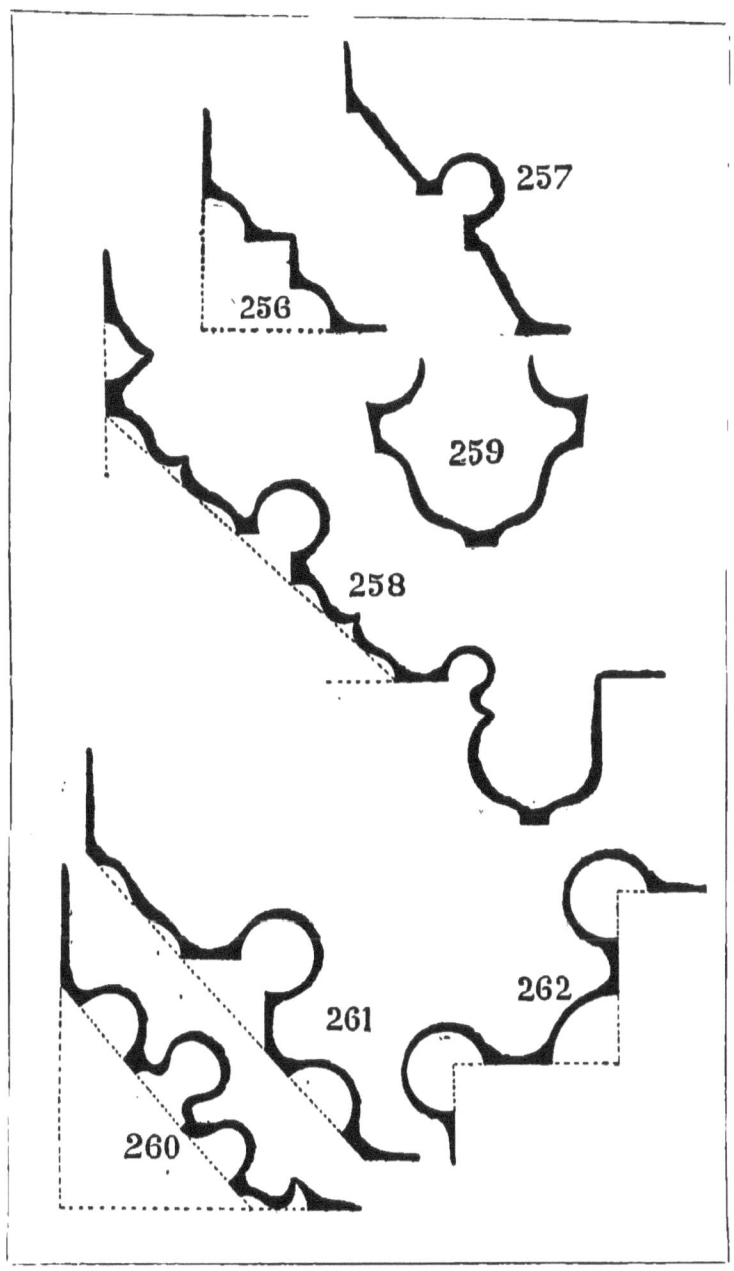

DECORATED.—256. Hollow chamfer. 257. Sunk chamfer. 258. Double ogee. PERPENDICULAR.—259 and 260. Roll and fillet. DECORATED.—261. Wave moulding, with part of chamfer plane left on. 262. Hollow chamfer, divided by ¾ hollows.

Perpendicular mouldings.—The mouldings peculiar to this period are the *casement*, the *bowtell* or ¾ *circle* used as a nook shaft, the *double ogee*, and the *roll and fillet*.

The mouldings are mostly arranged on the chamfer plane, and although the above-mentioned are peculiar to this period, most of the Decorated mouldings are also in use, but used with a decided tendency to flatness, and width as opposed to depth. This flat hollow is called a *casement*, and a common form is to find one or both ends of the

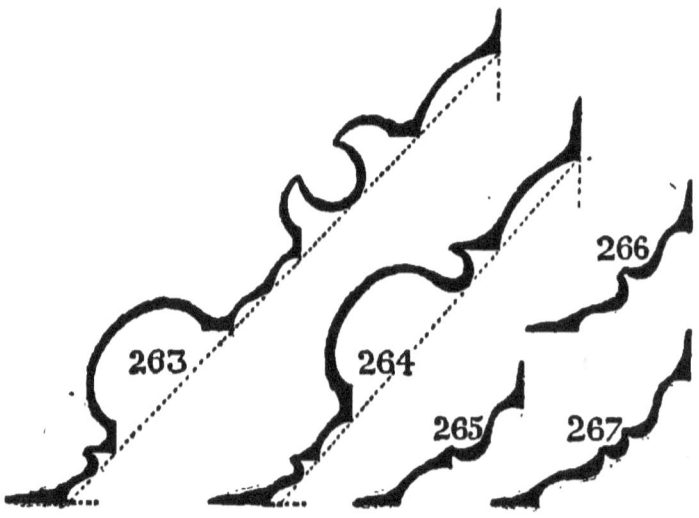

PERPENDICULAR.—263. The bowtell. 264. Casement. 265. Double ogee. 266. Double ogee. 267. Double ogee.

hollow returned in a kind of quasi-bowtell as seen in fig. 264. In later work this moulding occupies a large proportion of the width of the joint.

The *bowtell* formed as a shaft is a feature always to be observed in the style. Fig. 263 shows the usual position for such a shaft which is characteristic of the period.

The *double ogee* is much more common in Perpendicular mouldings than in Decorated. Figs. 265, 266, 267 show three very common varieties developed from the previous period. An ogee with a small bead or fillet at the base is shown in fig. 268, and fig. 269 is common.

The *roll and fillet* was not extensively used in the style, but when executed, took the forms in figs. 259, 260.

III. MODERN MOULDINGS.—In these days when mouldings have to be designed in harmony with the general style of the building they are to adorn, it is evident that a knowledge of mouldings in past periods is essential, not necessarily as a means of copying them into modern buildings, but as an educating factor in the designing of new ones. During

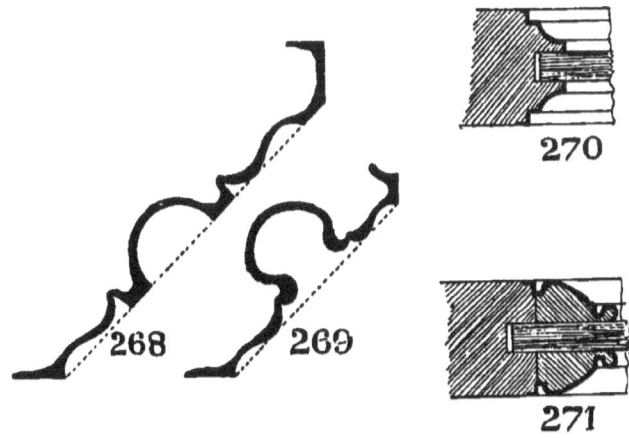

PERPENDICULAR.—268. Ogee with a small fillet. 269. Ogee with bead.
MODERN.—270 and 271. Panel mouldings.

the Renaissance period, the execution of mouldings in wood was carried to a great pitch of excellence, and work of the Elizabethan and Jacobean periods will be found very useful to the student in this direction. In the later periods also— in that of Inigo Jones and Wren, for example—the student will derive the greatest benefit from studying the best examples. We need only mention one example here, viz., the woodwork of the choir-stalls in St. Paul's Cathedral, by Grinling Gibbons, as one of the finest examples of this period.. South Kensington Museum is specially rich in work of the Renaissance period, in doors, chairs, tables, chests, and the like. The use of *bolection* mouldings may be first attributed to this period (see fig. 274). All sorts of combinations are used to produce the effect desired. Some examples of modern mouldings to doors or framing are given here (figs. 270, 271, 272, 273, 274). In these Grecian, and not Roman, models have served; the curves, with the exception of the beads, being sections of cones, show the

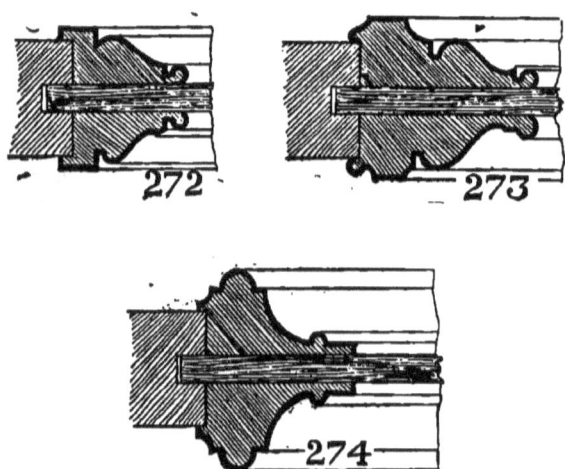

MODERN.—272 to 274. Panel mouldings.

infinite variety to be produced by the combination of very similar elements.

Lastly, the method of building up pieces of stuff on to a rough frame to form a series of continuous mouldings for such purposes as cornices, skirtings, architraves, &c., renders it essential that the stuff should be used in small widths, as the liability of wood to shrinkage will, unless care is taken, cause the pieces to split and fly. The method of framing these up is explained in the chapter on "Framing," and should be carefully detailed by the architect.

CHAPTER XVIII.

DOORS.

BEFORE going into the question of doors themselves, it will be well briefly to examine the methods that are usually employed in hanging them. Doors are generally either hung to—
1. Wrought-iron band hinges.
2. Solid frames.
3. Rebated linings.

Wrought-iron band hinges are only used for inferior work and for outbuildings. This method is shown in fig. 275. The hinges are usually at the top and bottom, and are hung on pins fixed in the wall, as shown.

Solid frames are generally used in the case of external doors, the position in the wall being altered according to circumstances. Fig. 276 shows, in isometrical projection, a frame of this kind for an external door. The feet of the two vertical posts are fixed in cast-iron shoes, having a projecting stud on the underside. These are fixed in the stone sill. The upper ends of the posts are tenoned into the head or lintel, which generally projects beyond the width of the frame, the projections, or "horns," as they are technically called, being built into the wall. A rebate is formed round the whole of the inside of the frame, into which the door shuts, the inner edge of the rebate being generally chamfered or beaded, as in fig. 277.

Rebated linings are mostly used for hanging internal doors, an example of which is shown in figs. 294, 295. This door is hung to the jamb lining itself, the latter being fixed in between the framed grounds.

Doors themselves may be divided as follows:—
1. Ledged doors.
2. Ledged and braced doors.
3. Ledged, braced, and framed doors.
4. Framed and panelled doors.

Ledged doors are the commonest form in use, and consist of V-jointed or beaded boarding, sometimes tongued to which are secured the ledges, on one side only. These are generally about 4 in. to 7 in. wide, and 1 in. in thickness, and are placed at the top, middle, and bottom of the door, as shown in fig. 278.

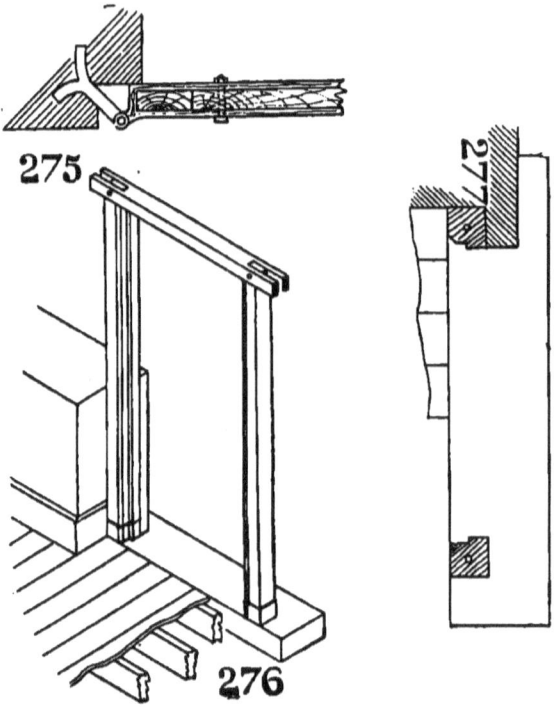

275. Wrought-iron band hinge. 276. Frame for external door. 277. Plan of frame for external door.

Ledged and braced doors are exactly the same as last described, except that the ledges are stiffened at the back by braces as shown by the dotted lines in fig. 278; these braces should slope downwards towards the jamb from which they are hung.

Framed, ledged, and braced doors are similar to ledged and braced doors, but they have, in addition, a framing of the same thickness as the top and bottom ledges, running up each side of the door, into which the ledges are tenoned. This is shown in figs. 279 and 280. The ledges in this form of door are generally called the top, middle or lock, and bottom rails, the two latter being wider than the top rail. The vertical parts of the framing are called *styles*. The styles and top rail, in ordinary domestic work, are

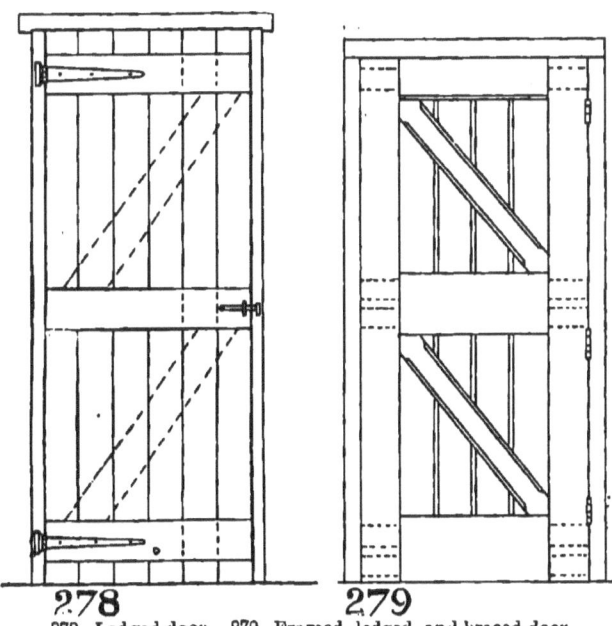

278. Ledged door. 279. Framed, ledged, and braced door.

usually 4½ in. wide; and the middle and bottom rail, 9 in or more.

In cases where the braces are left out the doors are called *framed and ledged*.

Framed and panelled doors are the most generally used, and are a step in advance of the last described in point of finish. Instead of the jointed boarding, panels are inserted in a groove, generally run in the centre of the inside edge of the framing. In order to diminish the width of these panels, a centre style or *munting* is framed in as shown in fig. 281, which shows a four-panelled door. Fig. 282 shows a six-panelled door, and the letters referring to their various parts are printed upon them. The plans of the various panels of these doors are shown in these figures, and the following are the technical descriptions by which each is identified, after stating the description, thickness, and number of panels:—

A. Stuck moulding.
B. Planted moulding.
C. Square framed and flat panel, both sides.

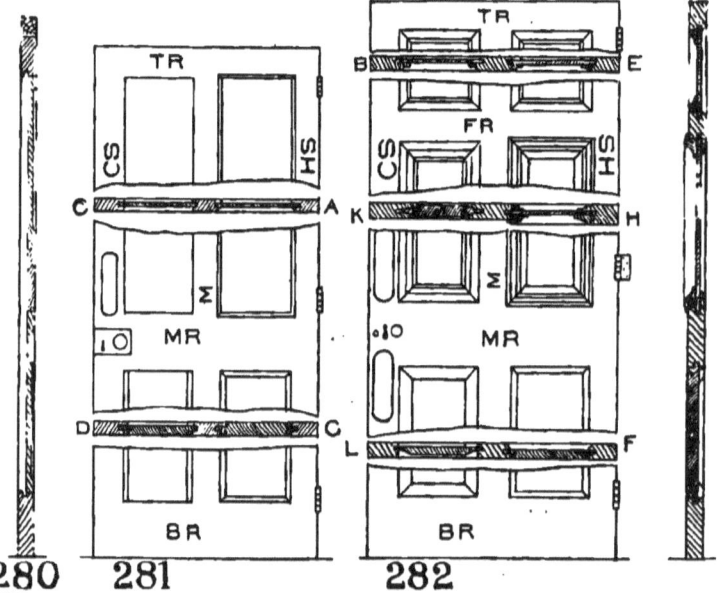

280. Section of framed, ledged, and braced door. 281. Framed or panelled door, four panels. 282. Six-panelled door.

D. Square framed flush panel, but bead butt one side.
E. Ovolo moulded, with chamfered and square back.
F. Bead flush, with stop chamfered back.
G. Bead flush front, with chamfered and flush back.
H. Bolection moulded one side, ovolo moulded back.
K. Moulded and raised panel, with moulded rising on each side.
L. Raised square panel in front, with square back.

The figures printed on the various parts represent the following:—CS are the closing styles, HS are the hanging styles. These are always in one piece, the horizontal rails of the door being framed into them. The internal faces of the styles have grooves down their centres about $\frac{1}{2}$ in. deep, and for about one-third of their width, to take the panels. M are the muntings, which are tenoned at both extremities into the rails, and are grooved on both sides as above

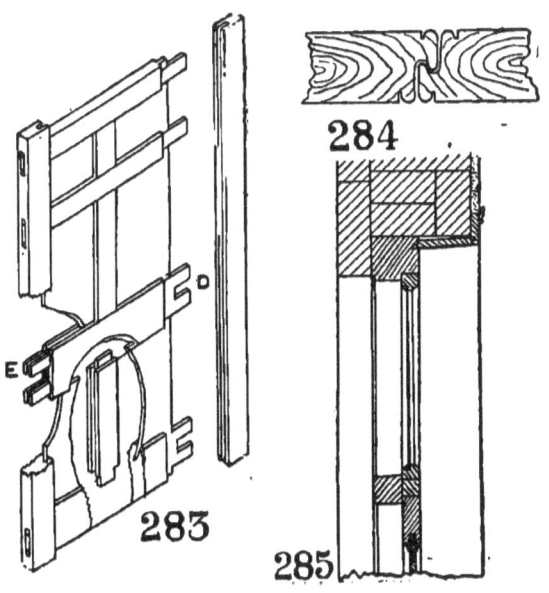

283. Method of framing up a door. 284. Junction of two meeting styles.
285. Section through fanlight, and portion of upper section of a door.

described to receive the panels. TR is the top rail; FR is the frieze rail; MR is the middle or lock rail, the centre of which is usually placed about 2 ft. 9 in. from the floor, as this is generally considered to be the best height for it. Many architects, however, prefer to place the lock rail somewhat higher, and 4 ft. is not at all an unusual height for the top of it. The lock rail should be tenoned into the hanging style with a double tenon having a haunch between them, as shown at D in fig. 283, owing to the width of the rail. At the other end, the lock rail should be tenoned into the closing style, with a double tenon, both in height and width, as shown at E in fig. 283, to allow of a mortise lock being inserted without cutting away the tenons. BR is the bottom rail, which should have a double tenon in height for the reason stated above. All rails should have a double tenon in width, when the door is more than 2 in. thick. Fig. 283 shows the method of framing up a door. It will be observed that the styles are longer than their actual finished sizes; this allows of the edges of the door being treated less *carefully*, and also prevents splitting when the wedges are driven home. These projecting pieces, or horns, are sawn off after the door has been glued up and cleaned off ready for fixing.

When the various parts of the door have been made, they should be carefully fitted together, and the door put on one side to dry until it is wanted for fixing; it is then taken to pieces and the joints cleaned. The whole door is then fitted together again, sometimes with the aid of a cramp, the tenons having received a coating of hot glue. The deal wedges which have been previously prepared are dipped in glue, and driven home into the mortises, their flat surface being against the tenons; the door is then left on a flat surface till the glue is dry, and the protruding portions of the wedges and the horns of the styles are then sawn off.

Dwarf doors are, as their name implies, of small dimensions, and are chiefly used for cupboards, cisterns, and other positions where required for special purposes.

Jib doors are those which are made exactly like a portion of a room, the chair rail, skirting, and other wall decorations being carried across them; in fact, the only indication of their existence is the line of the framing on the wall. They are seldom used, and are perhaps only justifiable to answer a special purpose in the decoration of a room.

Folding doors are those hung in two flaps, on each side of the door opening. Fig. 284 shows the junction of the two closing or *meeting styles*, as they are generally called in this case. The form of rebate shown is that generally used in the best work, as it is more adapted to prevent draughts, and, in the case of French casements, to exclude the damp.

Double-margined doors are hung in one flap, but by having a bead in the centre style, running from the top to the bottom, they are made to appear as if hung folding—*i.e.*, in two flaps. There seems no reason, artistic or otherwise, to justify their use.

A *fanlight* is a small casement fixed over a door for the admission of light into a hall or apartment. In this case the top of the door shuts against a *transom* or cross-piece, on the upper side of which the fanlight fits.

Fig. 285 shows the section through a fanlight and a portion of the upper panels of a door. The framing does not stop at the transom, but is carried right round the top of the fanlight, as is the architrave and all the other vertical linings and mouldings.

Sliding doors are those which are used when there is not room, or when it is not convenient, for a door to open on its hinges. The wheels on which the door is made to run may be fixed either at the top or bottom, though the former is generally found to answer best. The wheels are grooved on their circumference so as to run along an iron guide which keeps them in position. In confined spaces where there would not be room for one door to slide back and leave the opening free, two sliding doors are generally used, and these may or may not be in the same plane. Fig. 286 represents the sliding doors which are sometimes used to separate two drawing-rooms, for example. It will be seen that the doors are suspended by iron straps which are attached by jaws to the centre of the wheels, the latter running on iron guides. The wheels and guides are in the thickness of the soffit between the two rooms, and the doors themselves slide back in the wall which is constructed with a cavity down the centre to receive them. This is a far better method than the old style of folding doors which were so prevalent a few years ago, for these, owing to their numerous divisions and length, soon jammed and were unwieldy to manage. Fig. 287 shows the plan.

Mediæval doors in some cases showed the ingenuity of the joiner of the period, but it is impossible to do more than just to mention two, as examples. Fig. 288 shows the simple construction of a door from Staplehurst Church, Kent; the

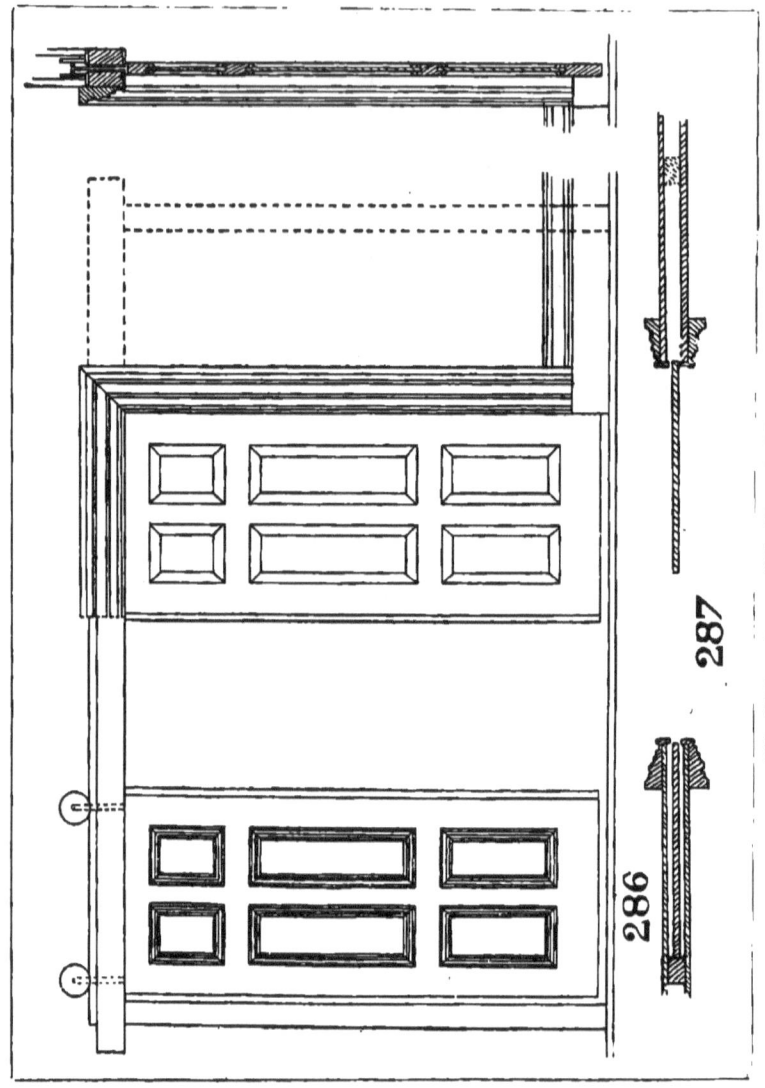

286. Sliding doors. 287. Plan of sliding doors.

CARPENTRY AND JOINERY. 197

door fits into a pointed arch, and the vertical tongued boards are kept in position by ledges. An effective framing of raised square panels is shown on plan in fig. 289, which

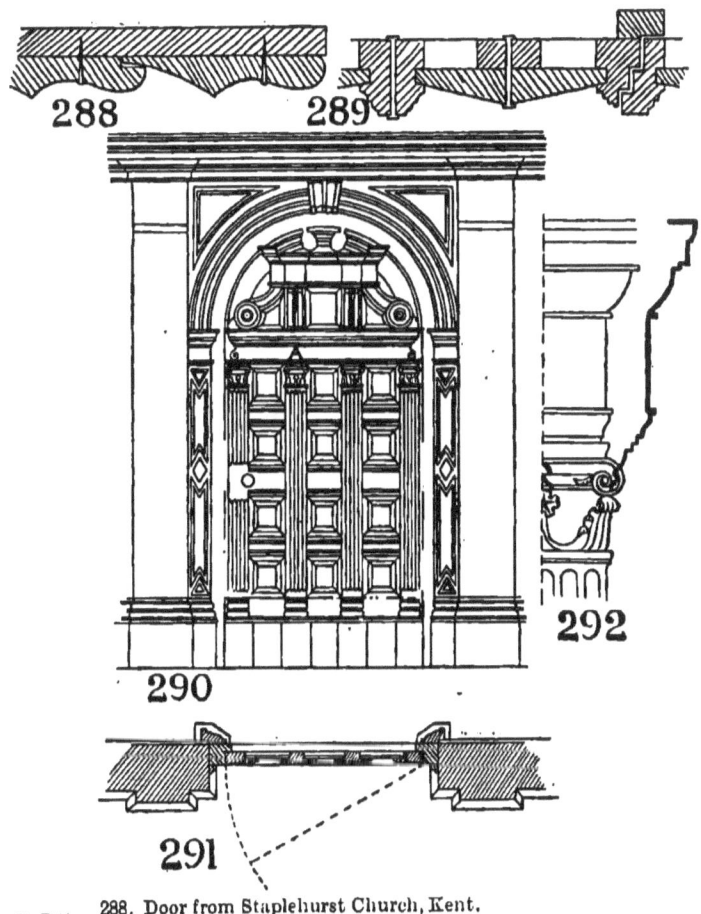

288. Door from Staplehurst Church, Kent.
289. Door from Holbeach Church.
290. Elevation of door to a billiard-room, Hyde Park Corner.
291. Plan of door to a billiard-room, Hyde Park Corner.
292. Detail at A, fig. 290.

illustrates the construction of a door from Holbeach Church.

An example of a modern door, of rather more elaborate

design than usual, is given in fig. 290, and was recently erected for a billiard-room in London. Fig. 291 shows the plan, and fig. 292 a detail at A.

Renaissance doors.—During the Jacobean period, doors of various types were constructed. A large proportion of the best examples of these are to be found in the South Kensington Museum. The method of construction is simple, yet effective, and they are well worthy of detail study by the student.

Finishings and fastenings must be left to individual taste, but the usual methods pursued will just be enumerated. Internal doors generally have linings or casings which are either left plain, or panelled, according to the thickness of the wall. They are usually plain for walls up to 14 in. in thickness.

On looking at figs. 294, 295, it will be seen that the jamb lining, JL, and soffit lining, SL, are fixed to the grounds, G, which are plugged to the wall; it will also be observed that the door itself hangs in the rebate of the linings, formed all round to receive it. This rebate also runs round the other external edge of the lining, which is therefore said to be *double rebated*. The architrave, A, which is placed round both sides of the opening, is also fixed to the grounds as shown, and mitred at the angles, fig. 293. Before the architrave is fixed, the plastering should have been completed, so that when it is fixed it will hide the junction of the plaster and the wood; this is a small matter which is much neglected, and consequently, after the plaster has dried, a fissure is often seen between the edge of the architrave and the wall paper.

The *furniture* of a door depends upon its situation and character. For the first three kinds of door mentioned in this chapter, a thumb or Norfolk latch and a rim lock are sufficient, as shown in fig. 278. For the better class of doors, mortise locks with ornamental handles are generally used in conjunction with finger-plates, a large one above the lock and a small one below.

The edge of the keyhole should be protected with a brass plate and escutcheon. External doors require to be further secured by horizontal barrel bolts; these should be vertical

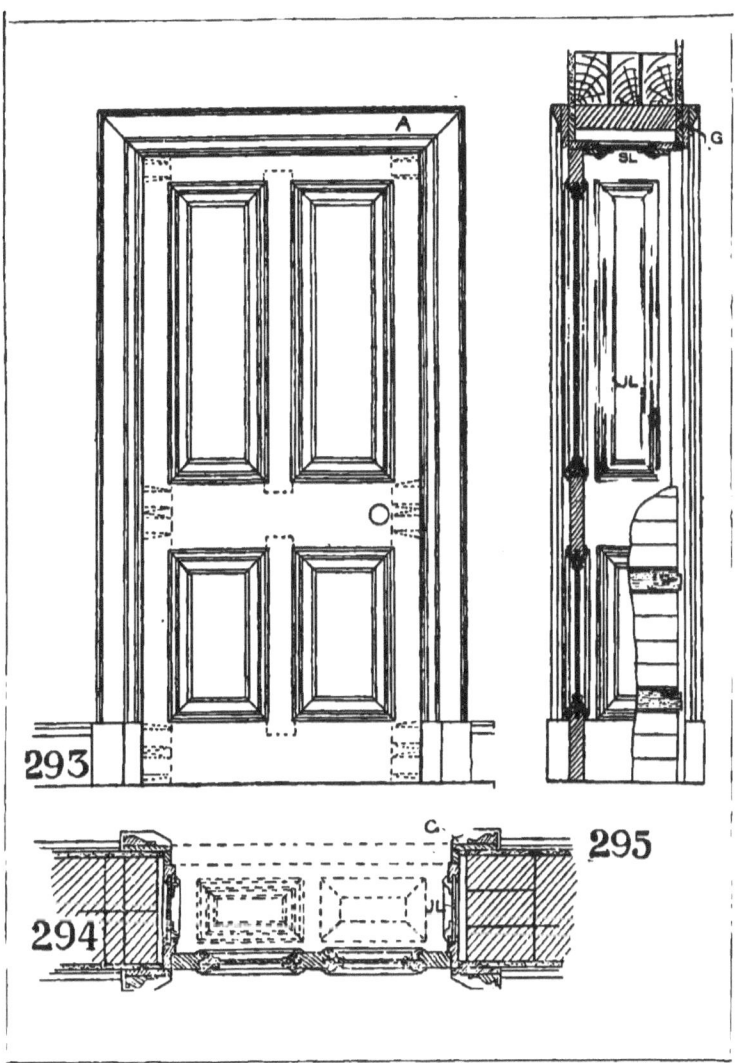

293. Internal door with rebated linings. 294. Plan of door with rebated linings. 295. Section of door with rebated linings.

when the doors are folding and should fasten into the head and sill. Chain and barrel fastenings are used in addition for front doors. There are numerous patent fastenings now in the market, such as fire-exit fastenings for theatres which release the door when pressed from the inside. Beside the band hinges mentioned at the commencement of this chapter, *cross garnet*, fig. 278, and *hook-and-eye* hinges are used for commoner doors. In framed doors, *butt* hinges are generally used, the upper fixed at the level of the lower edge of the top rail so as just to clear the tenon, the lower, for a similar reason, being placed just above the level of the bottom rail; the intermediate, if any, being placed half-way between the two. Further remarks on hinging will be found in the chapter on "Joints and Hinging."

CHAPTER XIX.

WINDOWS.

THE consideration of windows conveniently comes after that of doors. The designing of windows in relation to both the exterior and interior of a room is one of the most important duties of the architect. The window openings of a building should preserve the same character, and should, for structural reasons, be placed as far as possible from the quoins. Further, it should be remarked that where possible it is preferable not to have an even number of windows in an apartment because, if a pier is placed in the centre, it casts a shadow across the middle of the room, which is displeasing.

We may tabulate the various kinds of windows, with their frames, as executed by the joiner, as follows :—

    1. Solid frames with casements.
    2. Hollow, boxed, or cased frames, with sliding sashes.
    3. French casement windows.
    4. Window finishings, including shutters.
    5. Furniture and hinges.

1. SOLID FRAMES WITH CASEMENTS.—The solid window-frame (figs. 296, 297, 298) is made similarly to that for a door, and consists of two vertical posts, a head, and a sill. Round this frame is run a rebate, into which the casement shuts; this rebate is usually placed on the outside to enable the casement to open outwards, which is the more usual way, as it forms a more effectual barrier against the admission of rain. The sill, *os*, is generally made of oak, with its upper surface weathered to throw off the water, and throated so as to prevent the water being blown up and drawn in by capillary attraction. On the under-side of the oak sill is placed a metal *water-bar*, *wb*, which prevents any rain soaking in between it and the wall.

The casement (or sash, as it is sometimes called) is framed up with rails and styles in the ordinary manner, the space thus enclosed being divided—originally because glass could not be obtained in large pieces, and since continued for supposed effect—into small squares by means of sash bars, *sb*. In casements hung at their sides, the horizontal bars should be continuous from side to side in order to effectually withstand the jar of opening and shutting. These horizontal bars are merely mortised to receive the tenons

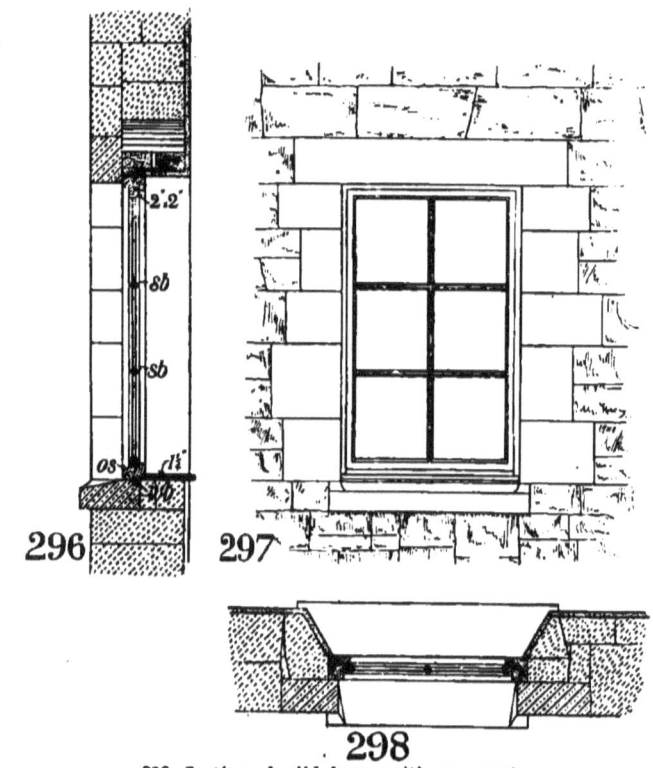

296. Section of solid frame with casements.
297. Elevation of solid frame with casements.
298. Plan of solid frame with casements.

formed on the ends of the portions of the vertical bars, which are scribed as shown, fig. 308, from *a* to *b*, to fit the moulding on the horizontal bar. The joint between the bars is sometimes effected by means of a mitre. The bars are rebated, as shown, to receive the glass.

A section in isometrical projection is given in fig. 301 of a casement hung to a solid frame in such a manner as to open inwards. If any water finds its way past the sash it falls into a groove which leads it back to the outside.

Fixed Casements are those which are screwed in position and are not wanted to open.

Casements Hung on Centres.—Such a casement is shown in figs. 299, 300. They are generally used in positions where they cannot easily be reached by hand, but can be opened by a cord. Pivots are fixed on the styles in their central axis, and these pivots fit into small iron sockets fixed in the frame to receive them. For the window to be opened, the upper portion is pulled inwards as that leaves less space for rain to drive in.

The method of making such a window weather-tight is by the following means:—A bead is fixed on the upper half of the outside of the frame, and on the lower half of the inside, as shown; then the casement, when shut, fits close against these, and by means of beads fixed on itself forms a continuous inside and outside bead.

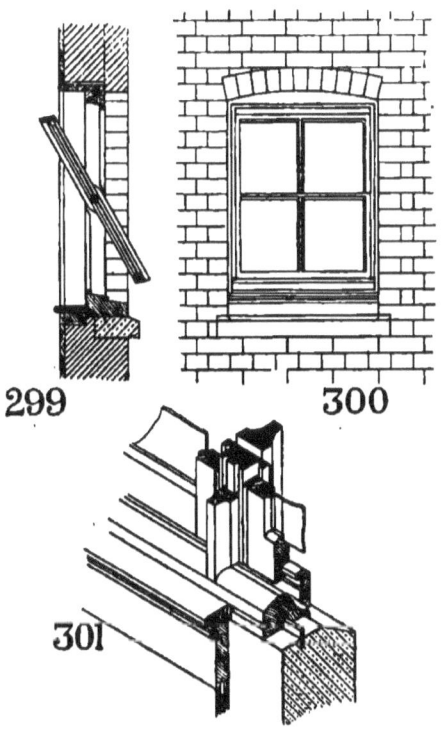

299. Section of casement hung on centres.
300. Elevation of casement hung on centres.
301. Casement hung to a solid frame.

2. HOLLOW, BOXED, OR CASED FRAMES, WITH SLIDING SASHES.—In this type of window the frame is made up as

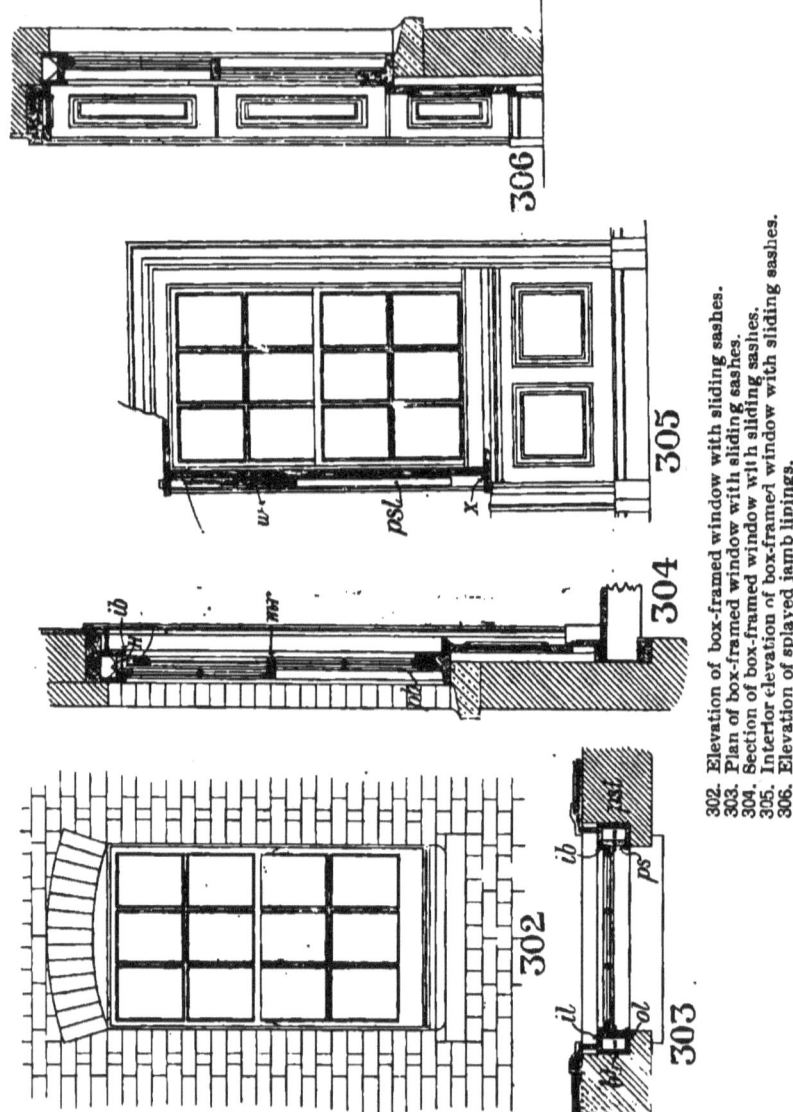

302. Elevation of box-framed window with sliding sashes.
303. Plan of box-framed window with sliding sashes.
304. Section of box-framed window with sliding sashes.
305. Interior elevation of box-framed window with sliding sashes.
306. Elevation of splayed jamb linings.

a hollow box or "case," made to receive the weights which counterbalance the sashes. Figs. 302, 303, 304 show such a window in plan, elevation, and section, placed in a 14-in. wall, but with 9-in. window back, the finishings extending to the floor. The frame consists of the inside lining, *il*, the back lining, *bl*, the outside lining, *ol*, and the pulley style, *ps*, so called because it carries the pulleys over which run the sash-lines supporting the weights. The various parts of the boxing are grooved together as shown in the illustration. The pulley style is grooved into the head of the sash-frame, *H*, and housed into the sill at the bottom, and secured by a horizontal wedge, *X*, as shown in fig. 305.

The upper and lower sashes slide in their places by means of a *parting bead*, *pb*, which is grooved into the pulley style, and housed into the oak sill. The inside bead, *ib*, is fixed with screws, so as to enable the sash to be removed when required for repairs. At the bottom of the window the inside bead should be made a few inches deep, as shown on section, in order to enable the lower sash to lift and admit air at the centre of the window without draught. The rails, where the upper and lower sashes meet, are called the meeting rails, *mr*, and the junction is formed by making them wider than the styles of the window by the width of the parting bead; and they fit tight to each other by a rebate, as shown, or by being simply bevelled. The lower rail of the lower sash is made from 4 in. to 6 in. deep, and is somewhat throated to keep out the weather, as shown (fig. 309). The method of hanging these sashes is as follows:—The styles of the sash have grooves taken out of their sides for a length of 6 in. or so from the top, into these the sash-lines or ropes are placed, and are passed over brass axle pulleys, *bp*, fixed into the pulley styles near the top, and having weights of lead or iron, *w*, attached, which counterbalance the weight of the sashes. Instead of rope-lines, which are liable to rot or otherwise get out of order, steel tape has of late years been introduced by patentees, and has the advantage of running smoothly and not wearing out. The weights are separated in the boxed frames by means of *parting slips*, *psl*, of wood or zinc, the upper end being passed through the head of the frame, as shown in fig. 305, and secured by a wood pin or nail. At the foot of the pulley style is formed a *pocket*, by means of which the weights are admitted, and which is covered by a flush piece, known as the *pocket piece*, the lower end being rebated and the upper rebated and undercut to the pulley style. The head of the frame is generally strengthened by blocks glued to the angle as shown.

The wall behind the arch over the window-frame is carried, as shown, by wood lintels and a brick relieving arch, or, as is very generally the case in these days, a coke-breeze concrete beam is used instead. The upper end of the frame can be nailed to this, and the sides are also nailed obliquely through the inside lining to the wood bricks or slips in the reveal. The tops and sides are further secured by wedges driven between the back linings and the masonry. If the frame is built in as the work proceeds, the head should be left longer than the width enclosed by the styles, so as to form horns resting on the masonry, as already pointed out in solid door-frames. Mention should be made of the various inventions which have been made towards improving the ordinary double-hung sash. In many of these the sashes are arranged so as to easily take out in the room for the purposes of cleaning, thus minimising the chance of accidents. Other patents have been on view at recent building exhibitions, such as the sashes balancing each other without weights, the upper one being made to open by itself as a casement sash at any inclination. They all deserve study as improvements on the useful but time-worn method of sliding sashes.

3. FRENCH CASEMENT WINDOWS have solid frames and casements hinged vertically, opening like a door. Figs. 312, 313, 314 show a French casement extending to the floor and becoming in fact a double door. In this case it is made to open inwards, the lower rail being fitted with a water-bar and having a throated sill as shown. These forms of windows generally admit the weather, unless extreme care is taken in their construction. It will be observed that a semicircular sinking is run along the rebate of the frame, so that if any rain should penetrate between

the casement and frame, it may immediately find its way out, and not be drawn inwards by any capillary attraction.

4. WINDOW FINISHINGS (including shutters).—Finishings, which include grounds, architraves, window boards,

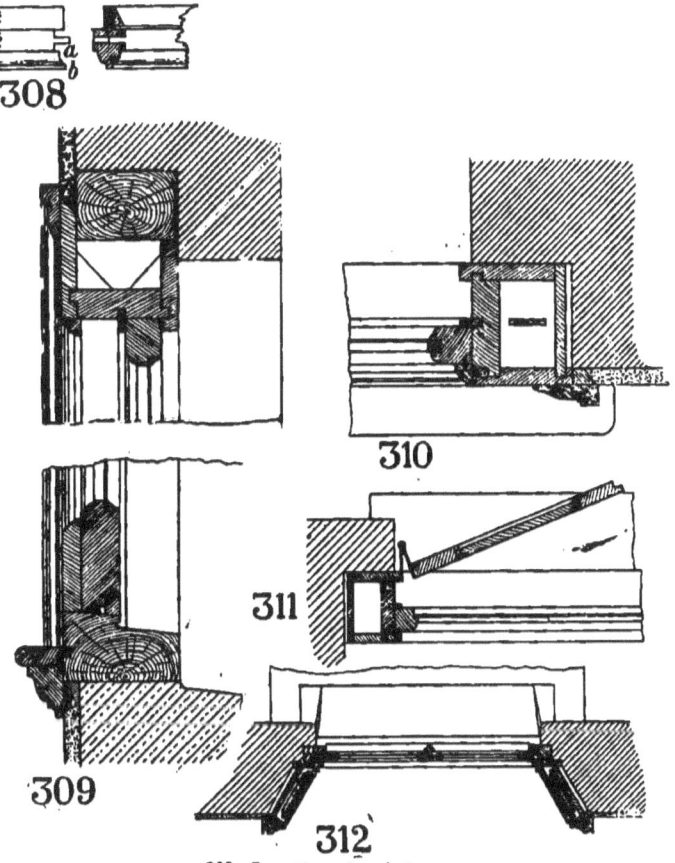

308. Junction of sash-bars.
309. Detail of sliding sashes in 9-in. wall.
310. Detail of sliding sashes in 9-in. wall.
311. Outside shutter.
312. Plan of French casement.

window back, &c., are the same for solid as for cased frames, so that they may be conveniently discussed under one heading. In a 9-in. wall (figs. 309, 310), the back of the window-frame generally finishes flush with the face of the plastering. The only finishing required in such a simple case is a window board, which may be moulded, and which is tongued and grooved into the oak sill in the same manner as shown in the illustration. In addition an

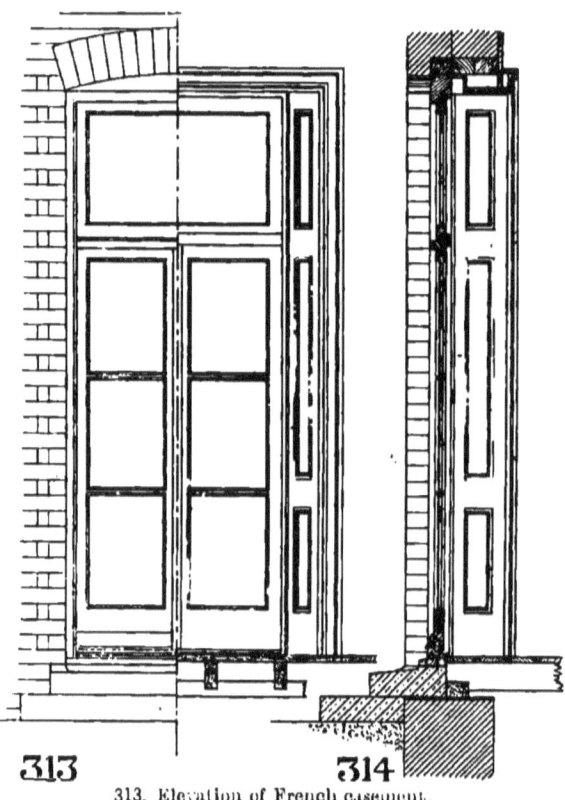

313. Elevation of French casement.
314. Section of French casement.

architrave is carried up the jambs and across the head, covering the joint between the grounds and plastering.

In a 14-in. or thicker wall, jamb linings have to be employed, being grooved into the inside lining of the frame, as shown in figs. 306, 307, and the window board is widened in consequence. The linings are further secured to *backings* plugged to the wall. The window back may be either panelled, as shown, of 1½ in. framing, moulded one side and fixed to grounds at the back. Jamb linings are frequently designed to be placed on the splay, in which case they are called splayed linings, and are fixed as shown in fig. 307. The architraves are often fitted in the best work with plinth blocks, as shown in fig. 305, which forms a finish against which the skirting may abut.

Linings over 11 in. in width should be panelled, and may be executed as shown in fig. 306.

There is one point in connection with window finishings which is often neglected, and that is, to provide a proper space for the blinds; this may be effected by lowering the architrave at the head of the window, forming a space behind for the blind.

In conjunction with window finishings, shutters deserve, perhaps, the most important consideration, although they are not, especially in towns, used so much as formerly. Shutters are two kinds, folding and lifting.

Folding shutters are so called because they fold back in one or more leaves into the boxings left for them on the jambs of the walls. In fig. 315, shutters are shown on plan to a splayed face. A space for the blinds to roll up and down is left by means of the piece of stuff BS; the face of this is continued, and has tongued on to it the back linings of the boxing, which is panelled and moulded, and fixed to backings like a jamb lining, which it really is when the shutters are shut at night-time. The inside end of this lining is tongued in a ground plugged to the brickwork, and placed to receive it. This has also a rebate for the edge of the front shutter, and serves as a groundwork for the moulded architrave. In regard to the shutters themselves, they may be treated either richly or plainly, according to the architect's designs. They are usually out of stuff 1¼ in. or 1½ in. thick. The face of the shutter exposed in the day-time is moulded to match the surrounding woodwork, the back part being left square-framed or bead butt. The edges are all rebated as shown, so as to form one continuous face when open. The design of the shutters in regard to general treatment and mouldings should be carefully set out in conjunction with the side of the room in which they are situate, so that the design may be continuous. Figs. 318 and 322 show the treatment to be adopted for the opening and closing of shutters so as to clear the blind space, which

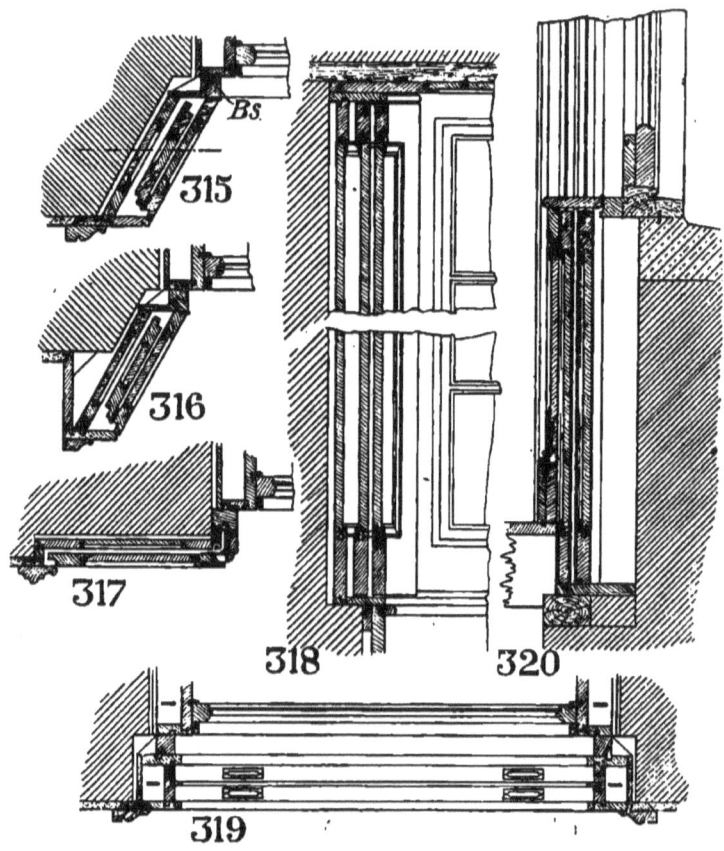

315. Plan of folding shutters.
316. Plan of folding shutters in a thin wall.
317. Plan of folding shutters in a thin wall.
318. Section of folding shutters (fig. 315).
319. Plan of lifting shutters.
320. Section of lifting shutters.

should be carefully considered, and to keep the shutters clear of the window-sill.

Folding shutters in thin walls are sometimes arranged as

shown in figs. 316 and 317, in which it will be seen that the boxings are blocked out beyond the face of the walls or are laid out flat along the face in order to obtain the necessary space.

There are other methods which will present themselves to the student, but which need not be referred to here.

Folding shutters may be hung in two heights for high windows.

Lifting shutters are sometimes used and have the advantage of less complication than the folding type. Figs. 319, 320 will explain their construction. It will be

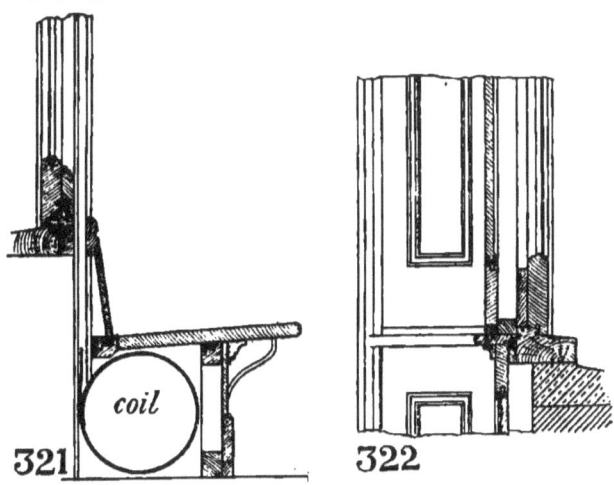

321. Shutter coiled under seat.
322. Section of folding shutter (fig. 315), closed.

noticed that the shutters are of framing the whole width of the window, enclosed and hung in the same manner as a double-hung sash, lead weights being generally used instead of iron in consequence of the weight required.

During the day-time they are let down beneath the window board into a space prepared for them. The window board is hinged and opens as a flap for the purpose. The shutters are lifted by means of rings let into their upper rail and are fixed together by means of a screw connecting the two leaves.

The student must use his inventive faculties for every

special requirement: as an example of such a case fig. 321 is shown. Here a *revolving* shutter was designed to work underneath a seat which was formed in a window opening. Seats formed in this manner are always useful and are sometimes framed as cupboards.

External shutters (fig. 311) where used are formed generally of a framing, fitted in with louvre boards, and hung by means of parliament hinges, so as to clear the reveal when opened.

5. FURNITURE AND HINGES.—The upper and lower leaves of a sash are generally held together when shut by means of a sash-fastener or clip, of which there are many varieties. The lower sash should be provided with two sunk lifts fixed in the lower rail, and the upper sash with a pair of handles screwed to the under-side of the meeting rail. Casements in solid frames require fastenings to secure the casements when shut or open. The commonest form is a casement-stay in iron or gun-metal. This is pivoted on to the casement and provided with holes throughout its length, which fit at discretion on to a pin fixed upon the sill. There are various other kinds of casement-stays which need not be described here. Casements are held in position when shut by a cockspur fastening fitted to one casement and entering a groove in the other, which forces the two together when turned.

In French casements, what is known as the *Espagnolette* bolt is the only satisfactory fastening. It is fixed on the inside of one of the meeting styles, and consists of top and bottom bolts connected by a long rod, arranged so that by turning the handle in the centre each bolt is shut, and the two leaves brought closely to each other. Casements hung on their bottom rail can be made to fall inwards by means of a quadrant stay (so called from being a quarter of a circle), of which there are various types. They are commonly employed in hospitals and schools, and in buildings generally, which are fitted with large windows.

CHAPTER XX.

FRAMING.

THE distinctive difference between carpentry and joinery has already been mentioned, but it is well to bear in mind that whereas the strength of the work executed by the carpenter is mainly dependent on the skilful disposition of the various parts, the joiner relies for stability principally on well executed joints.

One of the most essential points to be remembered in designing framing is that alteration in the *length* of the fibres is very small indeed, especially in woods having a straight grain, but that on the other hand, lateral expansion and contraction, *i.e.*, shrinkage across the grain, is very considerable, more especially in soft woods.

In the formation of joints this must be especially taken into consideration and sufficient *play* allowed, as if the pieces are too much confined, defects are bound to arise, as when timber expands it exerts a very considerable force.

Whilst heat causes metals to expand, and cold causes them to contract, the exact reverse is found to be the case with timber, owing to the sap being condensed by the cold, and the heat (of course) acting in an exactly contrary manner reduces the size of the wood, or *stuff* as it is technically called.

As previously explained, in the chapter on "Joints used in Carpentry," dovetailed joints should not be used in that trade. But in joinery the shrinkage of the dovetails tends to make the joints fit more closely.

From a consideration of the above remarks it is evident that it is desirable to reduce the wood to narrow pieces, thus distributing its shrinking propensities, and to fix it, where possible, in such a manner as to allow of expansion and contraction.

The subject of framing will be treated under the following heads :—

1. Grounds.
2. Architraves, cornices, friezes, and picture rails
3. Skirtings and dados.
4. Linings and ceilings.
5. Partitions.
6. Shop fronts.
7. Columns.
8. Fixing and glueing-up.

1. GROUNDS are small pieces of wood nailed to plugs (which are slips of wood driven into the joints of the brick-

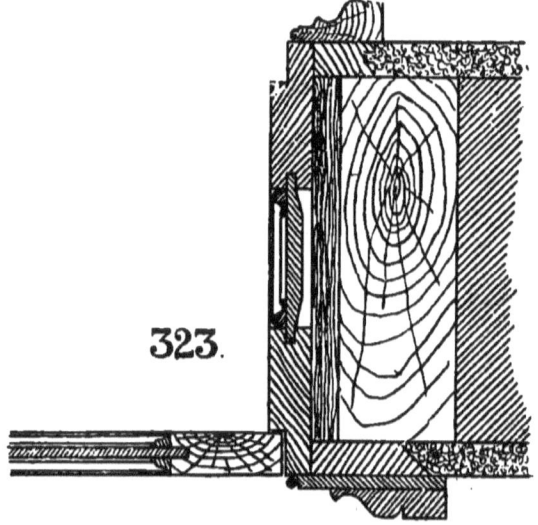

323. Fixing of grounds, architraves, and jamb linings.

work) or breeze bricks, so as to form a backing to which the joinery may be attached by nailing or screwing. As nearly all the apparent surfaces in joinery are fixed to grounds, and as their accuracy must depend upon the true fixing of these grounds, it is obvious that the latter require more care and attention than is usually bestowed upon them, and that their faces and edges should be fixed in true vertical or horizontal planes. Their thickness should be equal to the lathing and plastering of a partition (or the rendering, floating, and setting coats of plaster on a brick wall), so that when the latter is completed they may be in the same plane, and their edge towards the plastering should be splayed so as to form a key for the latter, which should be accurately worked up to this edge, which thus forms a solid finish (see fig. 323).

Grounds are left rough when they are covered by an architrave; on the other hand, when any part of their surface is exposed to view, that part is wrought, and generally has a moulding worked on it.

Framed grounds are used in openings for the better class of work, and form a rough kind of frame, with two upright posts mortised and tenoned into the head, though this junction is sometimes formed by simply notching or halving. Grounds should always be fixed in position before the plastering is commenced, as they thus form a "screed" or guide for the plasterer.

2. ARCHITRAVES (see fig. 323) are wrought borders, generally moulded and fixed round the openings of doors and windows to hide the joints between the *ground* and the plastering, and sometimes that between the ground and the lining. Architraves are often built up in two parts, glued together as shown in the above-mentioned figure, the base being merely a beaded board and the outer width moulded as desired.

Cornices are always better when framed together in lengths of small rectangular sections tongued together, though they are often made out of boards fixed at the rake of the cornice. They are fixed to grounds as shown in the upper part of fig. 326, having been previously glued and blocked together.

Friezes are not, as a rule, framed in themselves, but are tongued along their top edge into the cornice, as shown in fig. 326, and along their lower edge into the picture rail, which latter generally has a groove on its upper surface into which the brass picture hooks fit, as shown at A, fig. 326, and in detail, fig. 327. In an ordinary room they are generally fixed about 1 ft. 3 in. below the cornice when there is no frieze, and should be attached to grounds, as shown.

The height at which they should be placed is, of course, regulated by the proportions of the room.

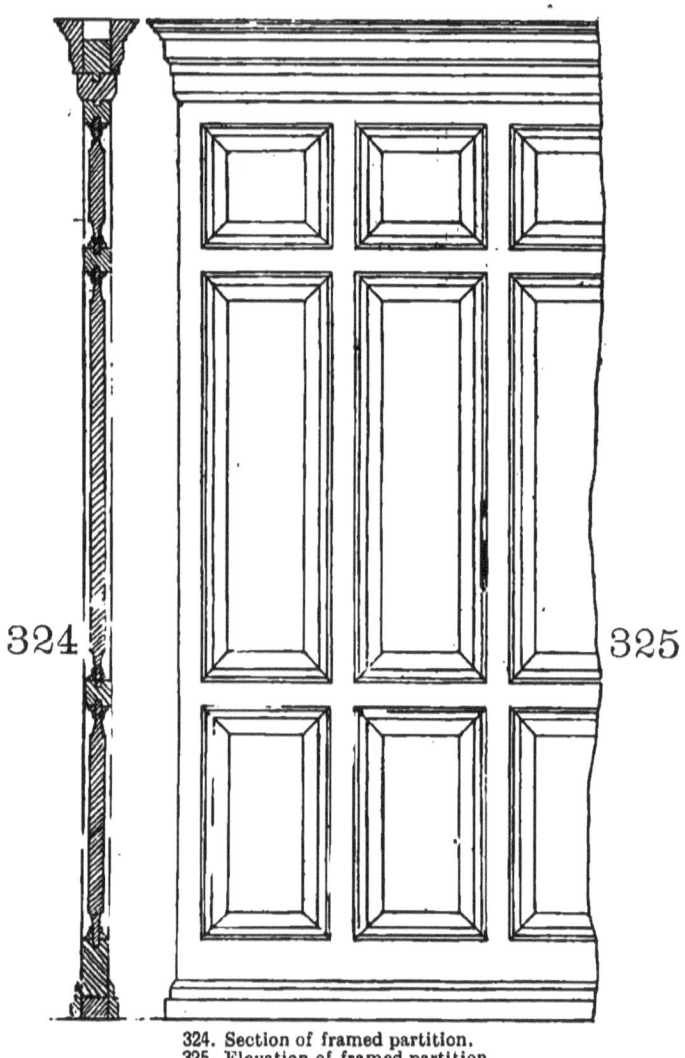

324. Section of framed partition.
325. Elevation of framed partition.

3. SKIRTINGS are intended to cover the junction of the

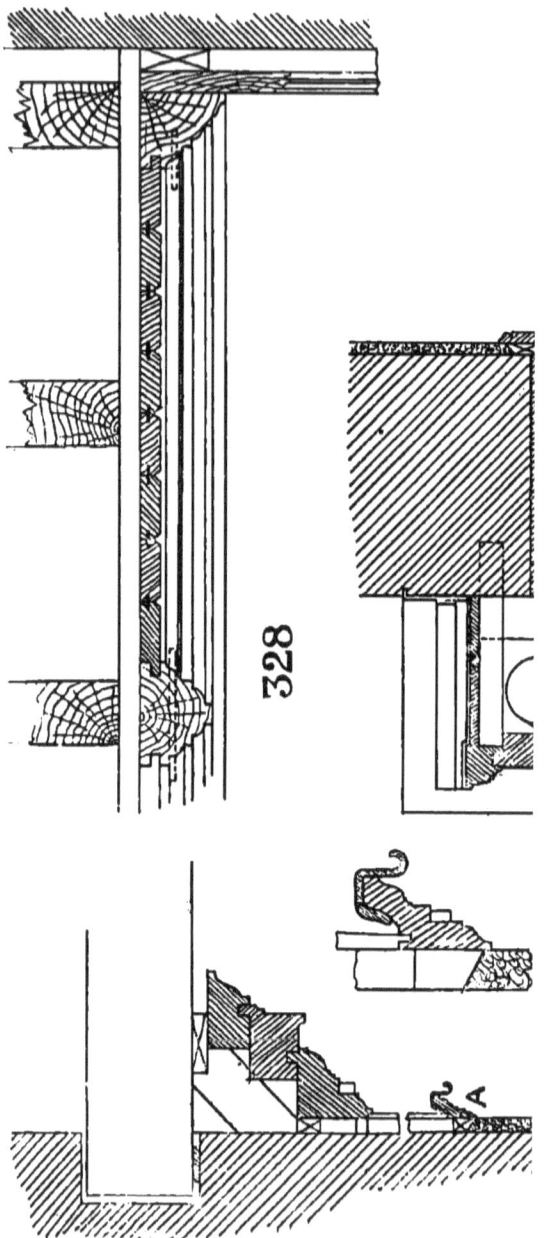

328

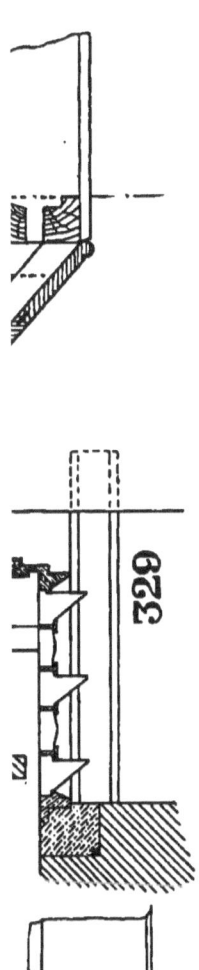

Fixing of grounds, architraves, and jamb linings. 327. Details at A, B, C, fig. 326. 328. Panelled ceiling. 329. Section of a shop front.

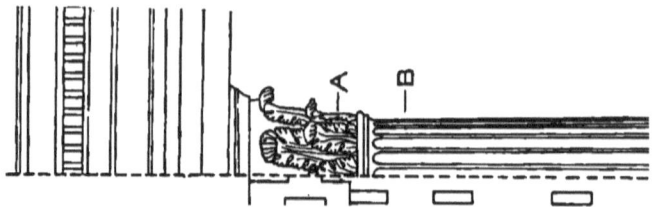

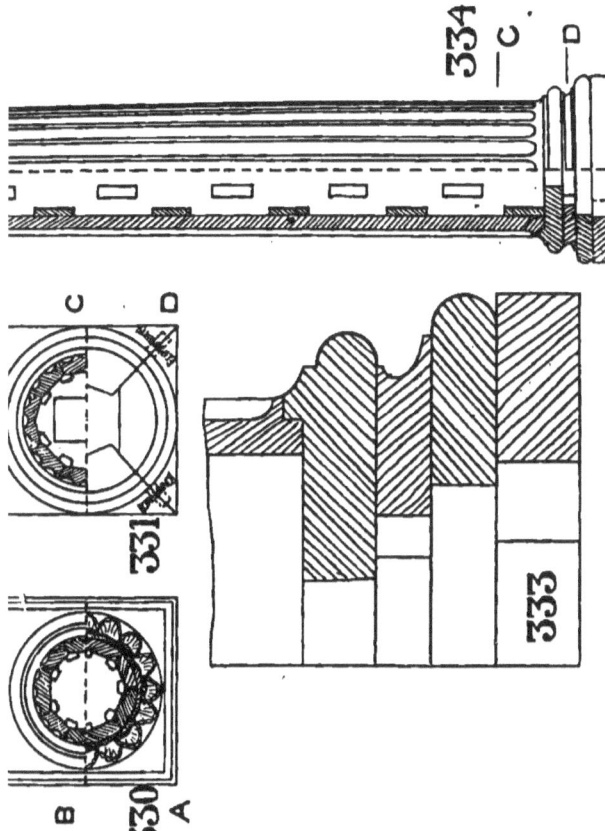

330. P'ans at A and B. 331. Plans at C and D. 332. Detail of glueing up of capital, fig. 334. 333. Detail of glueing up of base, fig. 334. 334. Half-elevation and half-section of column and entablature.

floor with the walls to protect the latter and to form a framing behind which the plastering may be secured.

They may be secured to plugs in the wall, but it is better to fix them to grounds when not attached to any joinery above. A narrow horizontal ground runs along behind the skirting near its upper part, to which the latter is fixed; it is also secured to a small fillet which is carried round the floor at its base. At C in fig. 326, and detail fig. 327, the skirting is shown as being made up in two pieces, which are tongued together, the lower member being also tongued into the floor; thus it will be seen that it is enabled to shrink at both joints without any fissure being seen.

Dados generally consist of panelled framing and may either be carried right down to the floor and be fixed against grounds in the wall at the base, and have a skirting attached in front, or they may be fixed as shown in fig. 326. In this case the dado is tongued on to the skirting and fixed at this point to a ground, the skirting being in two pieces. It will be seen that the grounds form a frame behind the dado, to which it is secured.

The top rail of the dado has a capping which conceals the junction of the plaster with the topmost ground.

4. LININGS are generally used to conceal some portion of the construction of doors or windows.

Jamb linings (see fig. 323) are those which are used in the vertical thickness of the wall (see figs. 294, 295, Chapter XVIII.)

Soffit linings carry out the same purpose in horizontal thickness (see fig. 295, Chapter XVIII.)

Window backs (see figs. 304, 305, Chapter XIX.) are those linings fixed between the skirting and the under-side of the window board.

Elbow linings are those that cover the splay in the wall between the sash-frame and the return of the wall (see fig. 306, Chapter XIX.)

Linings should be fixed as shown in the illustrations against grounds plugged to the walls, and placed usually about 2 ft. apart.

Ceilings are finished in wood in a variety of ways, the simplest of which is matchboarding nailed to the joists. Another method (fig. 328) frequently adopted in mediæval times, is to divide up the ceiling into a number of bays by moulded ribs placed in a parallel direction and attached to the joists above. These spaces are further divided into panels by transverse ribs, the edges of all being ploughed to receive the panelling, which, of course, is fixed in position first.

Coffered ceilings (see fig. 328) are formed by carrying the upper members of the cornice across the ceiling as ribs secured to grounds attached to the joists, the panel resting on the upper members of the rib; the intersection of the ribs is often covered by a carved boss, the panels themselves being frequently painted or carved.

5. PARTITIONS IN JOINERY are generally from $1\frac{1}{2}$ in. to 3 in. in thickness, wrought, ploughed, tongued, and panelled in various ways. Figs. 324, 325 illustrate a common form. This partition is secured at both of its vertical extremities. It will also be observed that it rests on a small sill, which is fixed to the floor, and that this joint is concealed by a narrow skirting. The framing throughout is of the same thickness, and the raised panels which are tongued into it are allowed to contract or expand at the joints in the same way as in the framing of a door, care being taken to fix the moulding to the framing and not to the panels. The cornice is fixed to the top rail, and is generally of the same section as the plaster cornice of the room.

6. SHOP FRONTS vary very much according to the taste and discretion of the architect, but fig. 329 gives a fair idea of the method usually employed. The cornice is framed up as before described, except that part of it generally comes away with the sun-blind as shown, and it resumes its position as part of the cornice when the blind is rolled up. A revolving shutter, which is composed of narrow strips of wood connected by hinged joints, is shown just below the sun-blind and behind the fascia. A continuous iron ventilating grating is shown above the sash, which is very necessary to prevent the condensation of the atmosphere against the glass, which would prevent the goods or articles on the stall-board being seen. This latter is too frequently neglected, sometimes no ventilation being provided at all, and at other times not an adequate quantity. The bottom of the sash is generally

fixed just above the stall-board, and a sill below this carries the whole sash between the supports and enables a bulkhead framing to be used, so that light may be obtained below. This is further assisted by the pavement lights.

7. COLUMNS.—Columns are framed together as illustrated in fig. 334, which is a section through an entablature and column with its base. Fig. 333 shows the method of framing up the base, the rectangular member being formed by mitreing together four pieces of "stuff" of equal length, which are secured by screws and glue, and are strengthened by blocks at the angles; when this has been done, the upper surface is truly planed, and the next course planted on and planed true, and so on until all the courses of the base are in position. When the block is completed, it is turned as required and rebated to receive the shaft. The latter is made of thin pieces of stuff, not more than 4 in. or 5 in. wide; these are jointed against wooden blocks as indicated, which should be placed in the centre of the joint. The pieces composing the shaft should if possible be made to the proper entasis before fixing. The framing of the capital (fig. 332) is sufficiently demonstrated, though very often these are made of thinner material, which is strengthened by wedges or blocks of wood glued at the internal joints (figs. 330 and 331 show plan at A, B, C, and D).

8. FIXING AND GLUEING-UP are very important points which must receive the careful attention of the joiner. In panelling it is generally found that wood of sufficient width cannot be obtained, more especially as no shakes or imperfections whatsoever should be allowed in the wood for the panels. If possible, the joints between the stuff forming the panel should be feather-tongued, so that if there is a shrinkage the light will not show through.

When plane surfaces have to be jointed, they should first be glued, and then, when dry, *keyed* at the back by the insertion of dovetail keys, fitted in similar or corresponding grooves across the joint.

Where great width is required, wide planks should not be used, but should be sawn up in pieces not more than 4 in. wide; these should be glued together, with the edge which grew nearest to the heart of the tree adjoining another farthest from it. Glued joints are often stronger than the wood itself, and their strength is not in proportion to the quantity but the quality of the glue.

When two pieces of wood are required to be jointed together to form a panel, &c., their edges should be carefully shot and then dried, after which they should be forced together so as to exclude all superfluous glutinous material. It should be remembered that when pieces of wood are joined end to end they absorb more glue than when their horizontal fibres adjoin each other, though in the latter case a better joint is made. In the case of veneering care should be taken to swell out the body of the material to be veneered with size, and thus, the veneer and the body being both swelled out, curling is obviated.

CHAPTER XXI.

SKYLIGHTS AND LANTERNS.

SKYLIGHTS and lanterns are formed in roofs in order to admit light into an apartment or staircase. They are naturally of varied forms to suit special requirements. When such a light is formed in the slope of the roof it is called a skylight, but when it is raised upon a vertical frame, filled in with sashes which form its sides, it is called a lantern.

SKYLIGHTS.—The most useful form is that in which the sash holding the glass is parallel to the side of the roof in which it is situated. An opening is formed by trimming between the common rafters, as shown in our illustration fig. 336. The space thus formed is lined with stuff, say 1½ in. to 2 in. thick, but more in case of a large opening; this lining projects 3 or 4 in. above the opening, and supports the sash, whose frame projects over it on all sides, the under-side being throated as shown to prevent water finding its way to the interior. Another method, just as frequently adopted, is to nail a cover-piece over the joint formed between the sash and the lining which supports it as shown in fig. 337.

A gutter is formed at the upper-side of the skylight where it joins the roof by means of a tilting fillet and a packing piece, as shown at G in fig. 336. A sheet of 6 lb. lead is worked round this and up under the slates as shown.

At the lower edge of the skylight the lead apron is carried up over the lining and formed into a small semicircular gutter as shown, in order to catch and intercept any water which may form on the under-side of the glass by condensation and prevent it running down the lining on to the ceiling. Too much stress cannot be laid on the importance of providing against water formed by condensation finding its way down on to the ceiling.

Condensation is sure to take place on any sheet of glass. It is caused by the heat of the room striking on the cold surface of the glass. Every precaution therefore must be made to carry off all such water formed by condensation. The sash-bars in a skylight are generally placed from 12 in. to 15 in. apart; they are continued down the slope of the

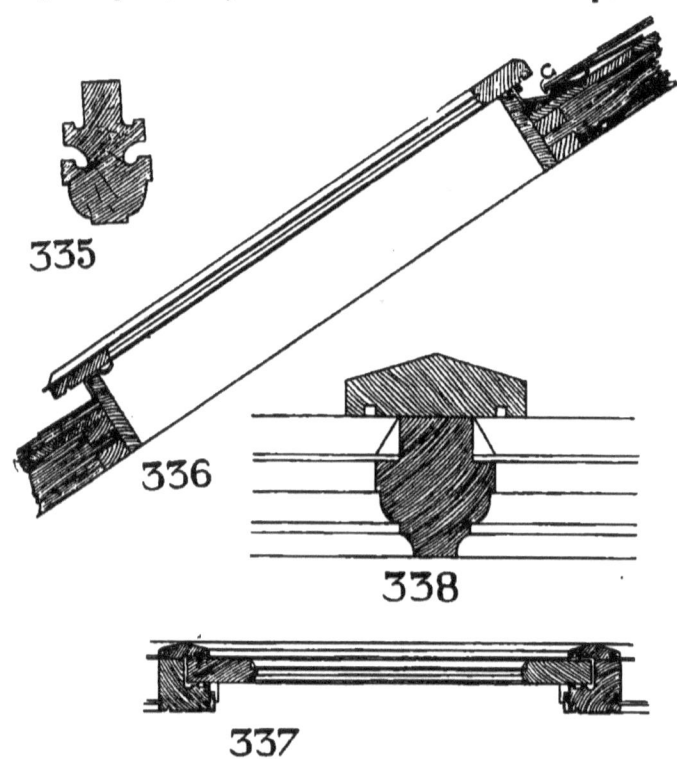

335. Detail of sash-bar. 336. Section through skylight.
337. Section through skylight made to open by sliding.
338. Detail of capping.

skylight without any interruption, and care should be taken to make them strong enough, not only to withstand wind pressure, but also the considerable weight of an occasional fall of snow. It is not an unusual thing to see sash-bars of skylights considerably bent by continual weight, and nothing is easier to prevent by making the bars sufficiently strong to carry the glass, which is usually heavy (rough plate), and to support any weight, such as snow, which may reasonably be expected to come on it. The sash-bars of a skylight have the same moulding as the styles and rails, and the same depth; they are usually from 2 in. to 3 in. wide, and framed at each end into the top and bottom rail.

The rebates should be grooved as shown in fig. 335, so as to carry off any condensed water or rain that might find its way through the putty. It will be noticed that the bottom rail is not as thick as the top one; this is done in order to allow the glass to be carried down over it and project beyond it over the edge of the bottom rail, so as to throw the rain well away from the junction of skylight and roof. In order to make a better joint against the introduction of rain, a capping is often fixed to the upper surface of the sash-bars, as in fig. 338. This is a very good arrangement, as it prevents any water penetrating through the putty or causing it to rot, as the rain which falls on the capping is thrown off on to the glass by means of the "throatings" formed on either side for the purpose. The rails of the framing are tenoned and wedged into the styles at the angles.

The glazing to skylights should be in as large sheets as possible, and should be continuous from top to bottom, it being remembered that there are no cross bars to intercept the flow of the water. When, however, glass cannot be obtained in sufficient length, and in consequence two pieces have to be joined, the junction is effected by means of metal clips (fig. 340), which being placed at intervals on the lower of two sheets has the upper one fitted into it. The sheet at the lower edge of the skylight is held in position by copper clips screwed at intervals to the rail underneath the glass, and clasping the latter at its lower edge.

Another form of skylight, or rather what may be termed a double skylight, is shown in fig. 339. It consists of a double pair of sashes similar to that shown in the last figure, but placed at the apex of a roof. It is raised above the level of the roof, which, in this instance, is of the queen-post type, by means of linings supported on the purlins resting on the queen-posts. The space between

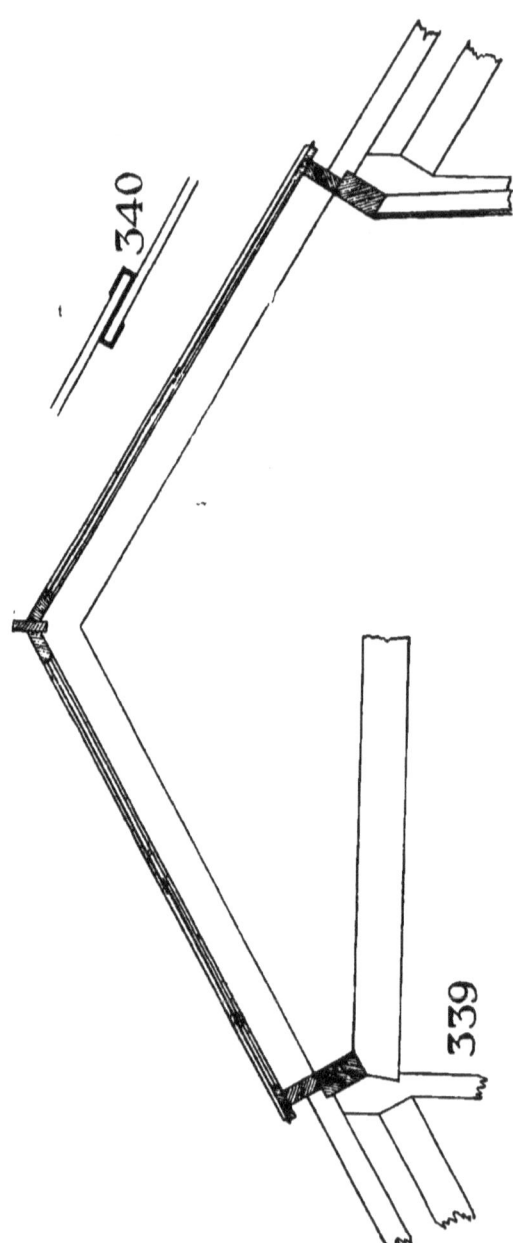

339. Double skylight 340. Detail of metal clip.

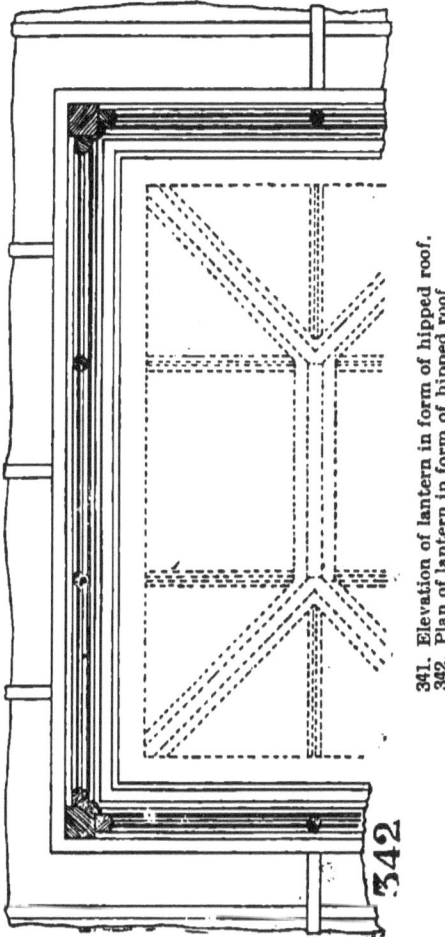

341. Elevation of lantern in form of hipped roof.
342. Plan of lantern in form of hipped roof.

the queen-posts is in this case lined with boarding so as to make it an inclosed shaft. The lower rail of the light should be grooved and made to fit on to an iron tongue formed on the lining, and a lead apron taken up the outside and passed into it.

A *laylight*, fitted as shown (figs. 344, 345), may be fixed at the ceiling-level, in order that the ceiling may be continuous. Laylights are discussed later on in this chapter.

LANTERNS.—As has been said, the term lantern is generally applied to any form of skylight or roof-light which is raised above the roof and has vertical sides. Figs. 341 and 342 show a lantern in the form of a hipped roof. It is 6 ft. by 4 ft. internally, and is formed so as to rise out of a flat roof. The opening in this case is trimmed with trimming joists, 11 in. by 6 in., and lined with stuff $\frac{1}{8}$ in. thick, framed and panelled.

The lantern light itself is made up of 4 in. by 4 in. angle-posts, into which are fitted 2 in. ovolo-moulded sashes, hinged at the top and opening outwards. The posts are tenoned into an oak sill, 8 in. by $3\frac{1}{2}$ in., which rests on a curb-piece whose scantling is 6 in. by 4 in. This sill has a condensation gutter formed on the inside as shown, and is further provided at intervals with *weep-holes*, leading any water which may accumulate to the exterior. The sill projects over the curb, and is throated so as to throw off any rain which may fall on it.

The lead apron lining is carried up under the oak sill and fitted into a groove, being held in position by means of a fillet. The roof in this case is formed at an angle of 30 deg. to the horizon, and is made up of $2\frac{1}{4}$ in. bars supporting $\frac{1}{4}$ in. rough plate glass. The glass, as mentioned for skylights, is in one continuous plane and held in position at the eaves by a copper clip fixed with brass screws. Where the sash projects over the head of the framing it is throated as shown.

The apex is crowned with a wooden roll covered with 6-lb. lead dressed over the top rail and held in position by clips at intervals. The meeting rails at their junction are further held in position by a cross-tongue.

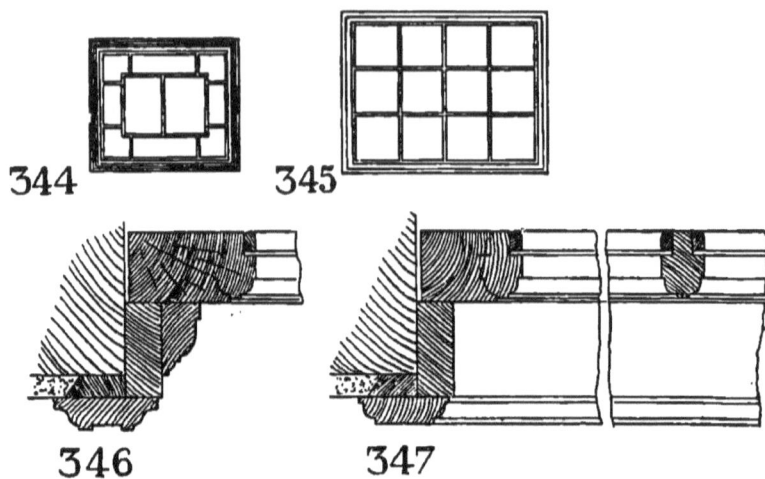

343. Lantern fixed at apex of queen-post roof. 344 and 345. Sketch plans of lay ligh 346 and 347. Details of lay lights.

In a lantern of any considerable length the side framing would have intermediate mullions, rebated for and filled in with casements in the same manner as the angle-posts to the present example. They would be of smaller scantling than the angle-posts, being probably out of stuff 4 in. by 3 in. nstead of 4 in. by 4 in. The angle-posts are, it will be observed, necessarily out of stuff the same scantling each way, as they have to line up with each face of the framing.

Another position in which a lantern is frequently fixed is at the apex of a king- or queen-post roof—an example at the head of the latter is shown at fig. 343. Here the lantern is raised by means of framed sides containing sashes,

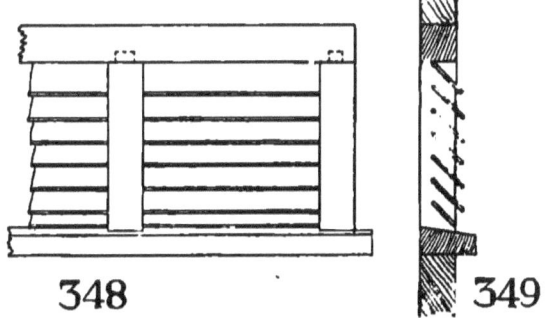

348. Elevation of glass louvres. 349. Section of glass louvres.

which are often hung on centres. The sill, which is formed of oak, rests on a purlin supported by the heads of the queen-posts. Similar precautions against water produced by condensation finding its way down into the roof must be taken, as in all the other instances. The heads of the two sides meet against ridge pieces as shown. In lanterns where continual ventilation is wanted, as in railway sheds, &c., the side framing may be fitted in with glass louvres (figs. 348, 349) instead of casements. These louvres, besides forming a continuous ventilation for hot or vitiated air and smoke, also admit the necessary light to the interior.

Only simple forms of lanterns have been treated up to this point, but it is evident that a great deal of elaboration can be adopted in their design. During the Mediæval period, many and various were the types of lanterns which were constructed for various purposes. Lanterns over staircases have been specially treated in the past. During the Queen Anne period and early in the present century examples were erected which bear evidence of thought and skill in design. The brothers Adam especially treated such features with great refinement and taste.

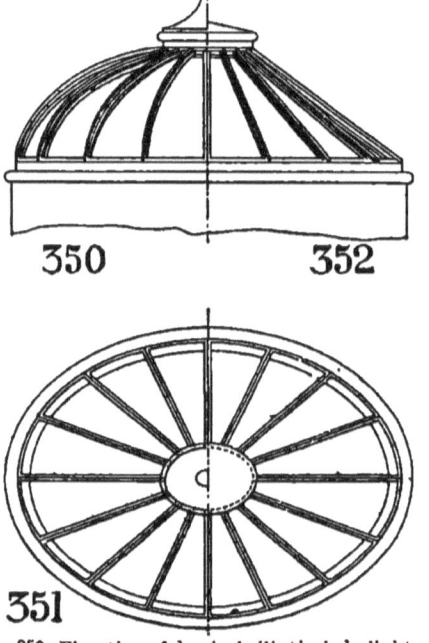

350. Elevation of domical elliptical skylight.
351. Plans of elliptical skylights.
352. Elevation of elliptical skylight.

Figures 350 and 351 show an example in plan and side elevation of an elliptical domical lantern of which the section on the minor axis is a circular segment. The method of setting out such a lantern will be described in a subsequent chapter. The ribs are all set out geometrically and filled in between with curved glass. Unless this form is particularly wanted it should not be employed, as curved glass if broken is difficult to supply and expensive, and no good is obtained by its use.

For this reason the shape shown in figs. 351 and 352 is preferable, and is, in fact, a very satisfactory form.

The example is an elliptical skylight in plan and side elevation. The construction is carried out in a similar manner to those described in the more simple forms. The crown of the lantern would nowadays often be fitted with an exhaust cowl to ventilate the staircase or apartment in which it is situate. Skylights or lanterns, when placed at the bottom of an area, should always be provided with a wire guard to protect them from anything falling through. Serious accidents, and even deaths, have been caused by failing to observe this precaution.

Laylights, sometimes called ceiling lights, are often formed at the ceiling-level for the sake of appearance, the light being admitted to these from an outer skylight often in the slope of the roof. These laylights are formed similarly as regards trimming, &c., to skylights, but there is less need to provide against condensation than in an outer light, although in certain cases it is advisable to make certain preparations for this. Figs. 344, 345, 346, and 347 are examples of laylights which have been recently constructed. The construction is simple, the space being trimmed; the laylight, formed out of 2-in. stuff, is made to rest on a stout lining nailed to the sides of the opening. Underneath is placed an architrave as shown, covering the joint of the plaster, ground, and lining. A sketch plan and detail is given of each laylight in order to show the general setting out.

CHAPTER XXII.

STAIRCASES.

It will be convenient to treat this subject under the following heads :—

1. Definitions and explanations of the various terms used.
2. The different kinds of wooden staircases generally in use.
3. The designing and construction of staircases.

1. DEFINITIONS of the various terms are given below, and the letters in brackets after each refer to the corresponding parts marked in the illustration.

The *staircase* is the name given to the space that contains the stairs.

A *tread* is the horizontal surface of the step upon which the foot rests when mounting or descending the stairs (T).

A *riser* is the vertical part of the step upon which the tread rests. The *going* is the horizontal difference between two risers (G).

The *rise* is the vertical distance between two adjacent treads.

A *nosing* is the front edge of the tread, usually moulded and projecting beyond the line of the riser beneath it (N). The line of nosing is an imaginary line connecting the edges of the nosing, which gives the angle of inclination of the stairs.

A *flier* is a step the enclosing lines of whose tread are all right angles (F).

A *winder* is a step whose boundary lines are not all right angles (W).

A *curtail step* is one in which the outer edge is projected or curved, so as to form a base for the support of the newel. The terminating scroll of the handrail is often placed in such a position in order that it may not interfere with the width of the stairs. Sometimes the lower two or three steps are of curtail shape (CS).

A *flight* is any number of steps without a landing. The *going of a flight* is the horizontal distance between the extreme risers of that flight.

A *landing* is a horizontal platform at the top of any flight.

A *quarter-space* is a landing half the width of the staircase.

A *half-space* is a landing the whole width of the staircase (HS).

A *newel* is a post receiving a handrail and forming a junction with flights of stairs and landings or other horizontal surfaces.

A *handrail* is a rail (generally moulded) parallel to the line of nosings at such a height as to aid people in ascending the stairs, viz., about 2 ft. 8 in. above the treads; on level surfaces, such as landings, it is generally placed 3 ft. above the floor. A *wreathed handrail* is one that is carried from the top to the bottom of the staircase without a break.

Balusters are vertical posts to support the handrail; they should not be more than 5 in. apart.

A *ramp* is a sudden concave curve in one direction, generally occurring in handrails.

A *knee*, as distinguished from a ramp, is a sudden convex curve.

A *swan's neck* is a combination of a ramp and knee. A *string* is a piece of wood placed at an inclination to support the treads and risers. A *wall string* is placed against the wall and fixed to it (WS). An *outer string* stretches from newel to newel, and supports the outer edges of the treads and risers. A *cut string* is one cut to the shape of the stairs, the treads and risers being simply fixed to the horizontal and vertical faces of the notches (see fig. 353). This form is only used for the commoner stairs. A *cut and mitred string* is of similar construction to the last, but the ends of the risers are mitred against the vertical notch, and a moulding is carried round the two exposed edges of the tread. The two last named are called *open strings*. *Close strings* have their upper and lower edges parallel, and fig. 354 shows the method adopted of housing the treads and risers. *Rough strings* or *carriages* are used between the outer and the wall strings when the stairs are more than about three feet

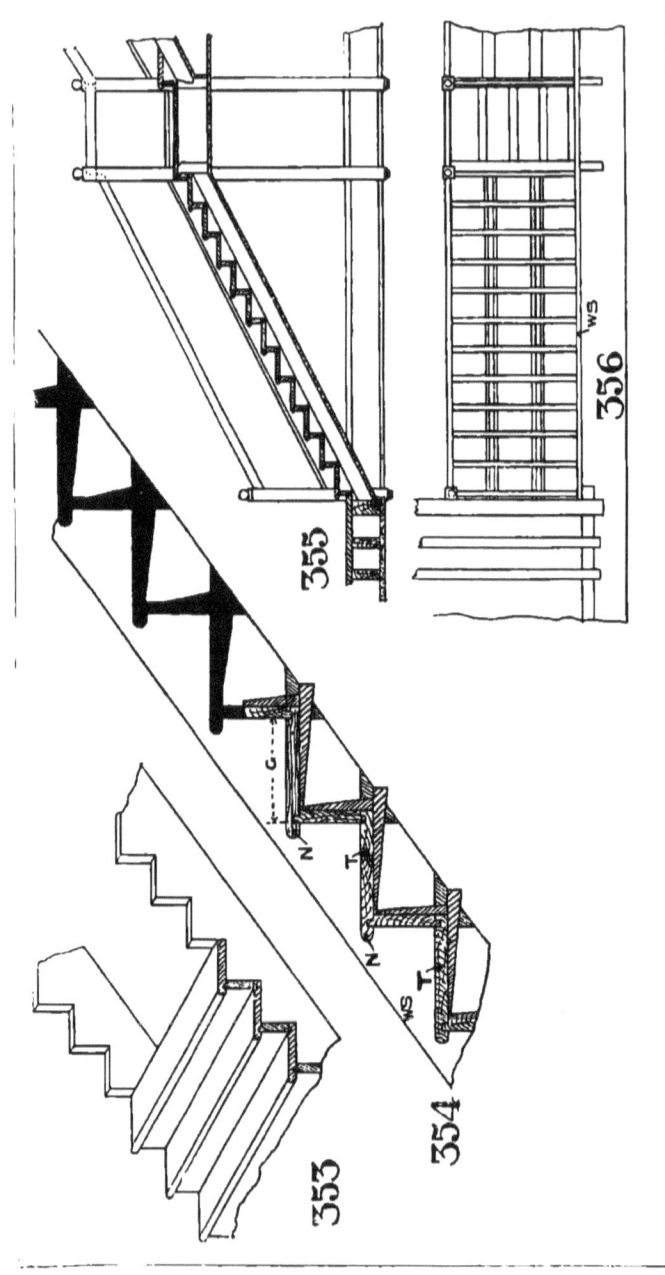

353. Cut-string stairs.   354. Close-string stairs.   355. Section of straight stairs.   356. Plan of straight stairs.

wide, and help to stiffen the treads and risers and prevent any tendency to bending.

The *spandrel* is the triangular space between the treads and risers and the floor of the lowest flight, and is generally enclosed with a panelled front called the *spandrel framing*.

A *well-hole* is the space contained by the handrails between flights of stairs going in different directions.

2. THE DIFFERENT KINDS OF STAIRCASES are as follow: —*Straight stairs*, which are used for very narrow staircases. If there is a wall on either side, the wall string is fitted against each, and the treads and risers secured to these. If there is but one wall, the upper end of the outer string is secured to a newel-post if there is a break in the flight, and trimmers are carried from the posts to the wall, as shown in figs. 355 and 356.

Dog-legged or newel stairs are those in which the width of the staircase (or enclosure in which the stairs are formed) is divided longitudinally into two equal widths, and in which there is no well-hole.

Open newel stairs have a well-hole between the flights around which the newels are arranged.

Geometrical stairs have the flights arranged round a well-hole, but have no newel-posts, each step is housed into the wall string and is also supported by the outer string. It is also upheld by the step below it, and therefore the treads should be of substantial scantling. The handrail is carried from top to bottom of the staircase without interruption. These stairs require more skill in construction, and were much used in the period of the Georgian era. They are not, however, as a rule, either so strong or so satisfactory as the open newel staircase.

Circular newel stairs consist of steps whose extremities are supported by the wall at one end and at the other by a newel or column from which they radiate to the wall.

Circular geometrical stairs radiate from an open well-hole in the centre, and have their other ends pinned into the wall, each step also derives some support from the step beneath it.

3. THE DESIGN AND CONSTRUCTION OF STAIRCASES are

Q

matters of great importance in the appearance and comfort of a house, the details of which are too often neglected. The design, of course, varies with the class of house, but the following points should always be observed. Stairs should consist of flights, running alternately in opposite directions. Such flights, as a rule, should not contain more than from ten to twelve steps, and they should not rise more than 8 ft. without a landing. The landings between the flights should be of a length and width not less than the length of the steps. Winders and isolated steps should be avoided as much as possible. The treads and risers should be proportioned as follows:—Multiply the rise in inches by 2, and add to this the width of the tread in inches; the result should be 23 in. From this it will be seen that the wider the tread the less will be the rise. Steps that have a wide tread and high rise are very fatiguing to ascend. Care should be taken that there is plenty of head room between the flights, and 7 ft. is generally considered the least that should be allowed for. Stairs should not, as a rule, be less than 3 ft. in width, so as to allow two people to pass. They should always be well lighted, more especially at the commencement and termination of a flight. A lantern light, at the top of an open newel staircase, is an ideal method. Where this is not possible light should be obtained on every landing.

Before planning a staircase the following details should be ascertained : (1) the position and sizes of all openings which the stairs must not interfere with ; (2) the possible width and length of the flights ; (3) the heights between the floor-levels, which should be marked on a story-rod; (4) the position of the lowest and the top riser, which, of course, must be kept away from all openings, &c. The height of the risers must then be determined, so that they will go exactly into each story, and these should be marked on each rod. The plan can then be laid out when it has been decided what description of stair is most desirable, whether dog-legged, open newel, &c., avoiding winders as much as possible.

With regard to the latter, it should be kept in mind that at a distance of about 1 ft. 6 in. from the handrail, they

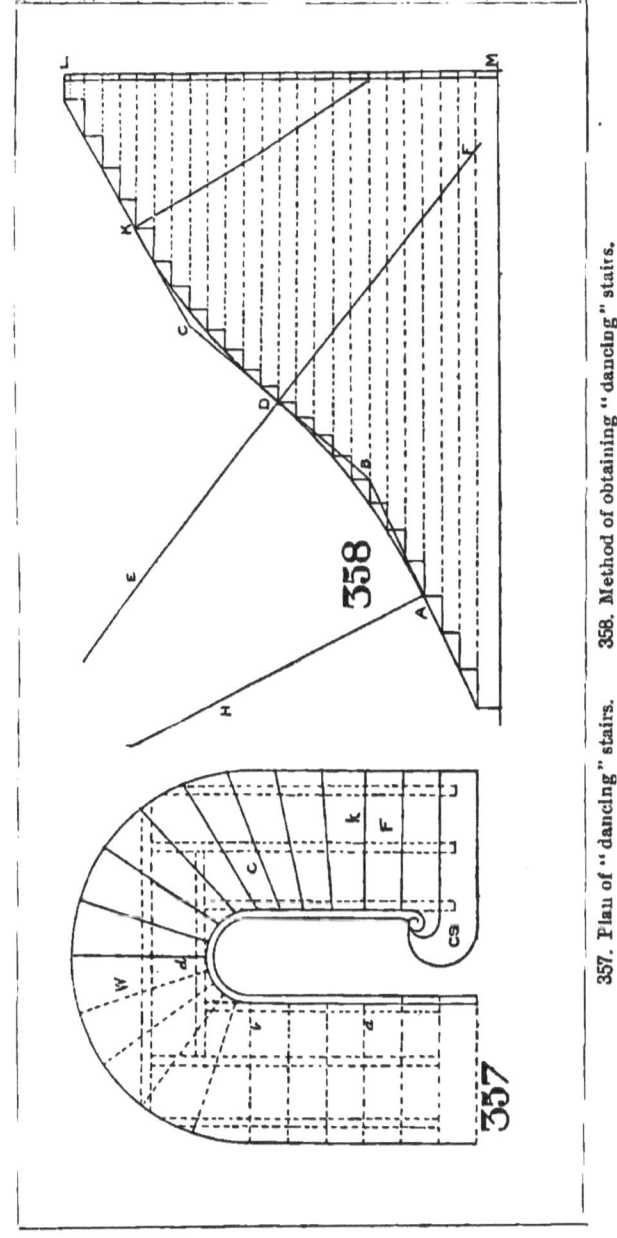

357. Plan of "dancing" stairs. 358. Method of obtaining "dancing" stairs.

should, if possible, be of the same width of tread as the fliers, because this is approximately the position of anyone ascending or descending the stairs when using the handrail. The distance from the handrail should be half the length of the treads when they are of a less width than 3 ft.

In the case of geometrical staircases with winders, they are often drawn converging to the centre of the semicircular termination of the handrail on plan, as shown by dotted lines in fig. 357. This, however, makes the treads very narrow near the handrail and very wide towards the wall, and as the inclination of the line of nosings of the winders is much steeper than that of the fliers, an awkward knee is given both to the string and to the handrail. To obviate this, the steps, with the usual exception of the first and last four, are made to "dance," as shown by firm lines on fig. 357, *i.e.*, the inequality is distributed amongst them as follows : Let AB, fig. 358, represent the line of nosings of the fliers, and BC that of the winders, as shown in dotted lines in fig. 358. Bisect BC in D by the straight line EF, make BA equal to BD, and set up AH at right angles to BA, till it meets EF produced. From this latter point of intersection describe the arc ABD. In like manner find the arc DCK. We have now a flat cyma without any knee. From the vertical height LM, draw the horizontal lines through to the curved line of development which will give the risers and treads, which latter can be transferred on to the plan.

The construction of stairs is illustrated in figs. 359 to 370. Figs. 359 and 360 show the plan and section of *dog-legged stairs*. In this case there is a half-space landing carried by a trimmer going from wall to wall, and to which the newel is fixed, the outer string in both cases being tenoned into the latter. Rough strings (RS) are framed between the trimmers, and rough brackets (RB) are nailed on to them to assist in supporting the treads and risers. A plan and transverse section of an *open newel staircase* is shown in figs. 361 and 362. It will be seen in this case that the rough string abuts on a pitching-piece (P), and that bearers (B) are carried across the quarter space into the wall so as to carry the risers for the winders, and these are sometimes stiffened

CARPENTRY AND JOINERY. 243

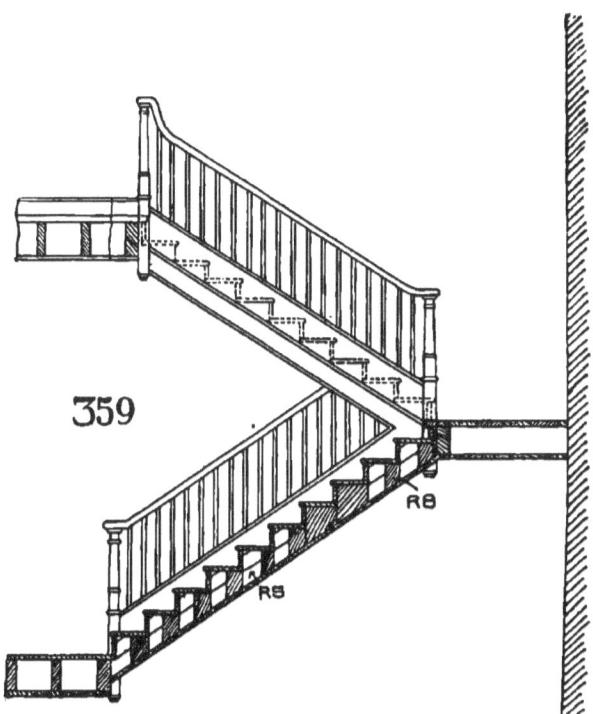

359. Section of dog-legged stairs.

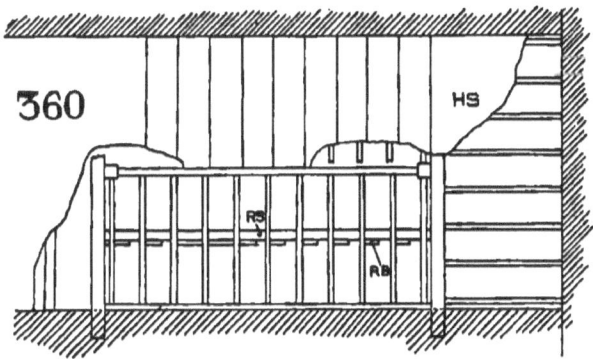

360. Plan of dog-legged stairs.

by cross bearers (CB). A trimmer is carried to the third newel, which is fixed to it, and the stairs start ascending again from this point. The outer string is housed into the newels as before, and small lengths of handrail and outer string connect the second and third newels.

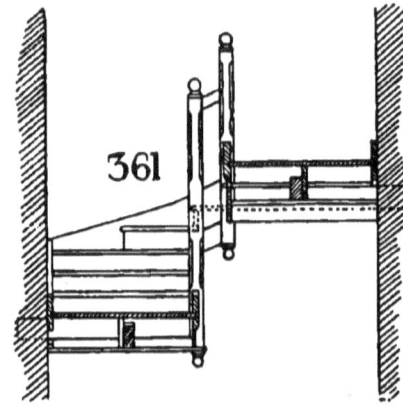

A *geometrical staircase* is shown in figs. 363 and 364 with dancing steps, both in part elevation and elliptical on plan. The position of the carriages is shown bolted thereon. The position of the principal carriages is determined in the following manner: — Lay a straight-edge on the plan and move it about until a straight line is found, which may be divided as nearly equally as may be by the intersections of the risers. The object of this will be manifest when it is remembered that, when the steps are of nearly equal width and rise, the angles will be approximately in a straight line, so that there are very few cases in which carriages may not be used for stairs if the above method be pursued. This method of

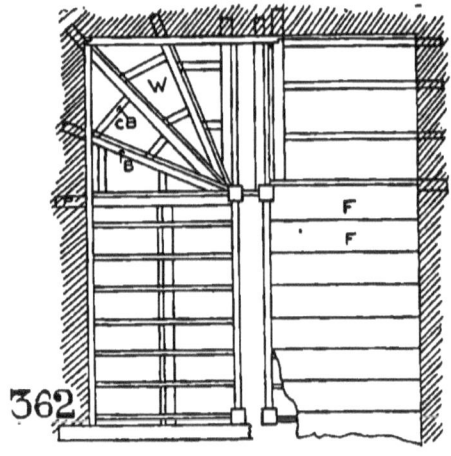

361. Section of open newel staircase.
362. Plan of open newel staircase.

CARPENTRY AND JOINERY. 245

forming the carriages of stairs was introduced by Mr. John Newlands, and is much stronger and more satisfactory than the older and more laborious method of framing for every

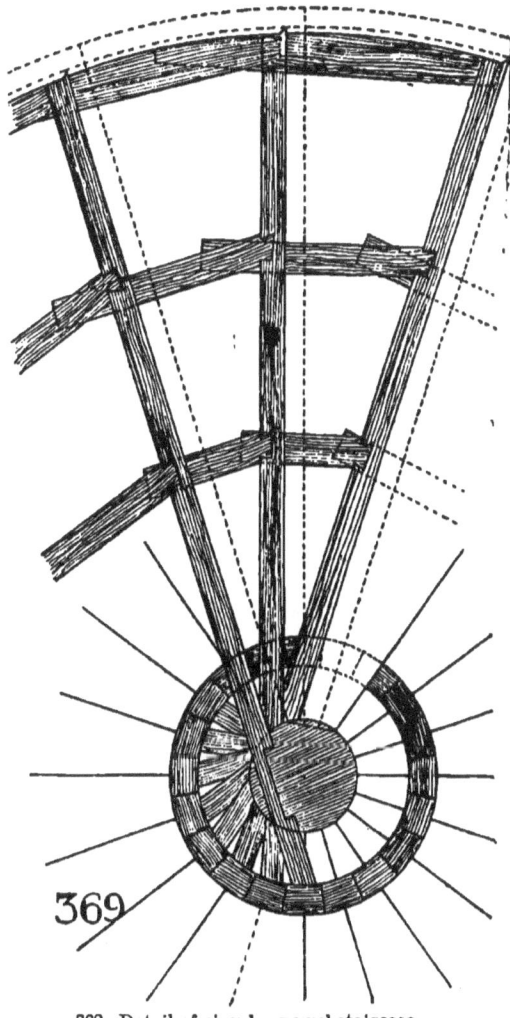

369. Detail of circular newel staircase.

step, and it has also the advantage of being less expensive. Figs. 365 and 366 show details. Figs. 367 and 368 show

246   CARPENTRY AND JOINERY.

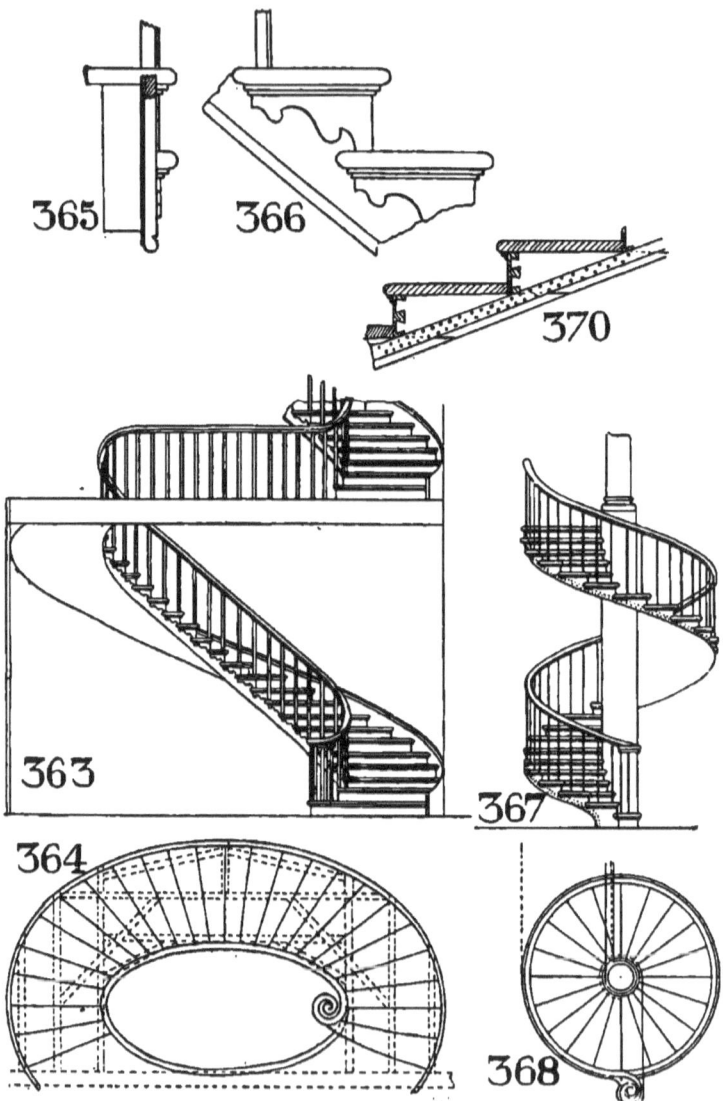

363. Elevation of geometrical staircase.   364. Plan of geometrical staircase.
365. Detail of geometrical stair.   366. Detail of geometrical stair.   367. Elevation of circular newel staircase.   368. Part plan of circular newel staircase.
370. Detail of circular newel staircase.

part plan and elevation of a *circular newel staircase*. The lower part of the newel is made up of staves of 2-in. planking, into which the risers and bearers to each step are fixed. It is best in this case to tongue the risers into the string, which should be in small pieces; a band of iron is then screwed round both, as shown in dotted lines on the elevation and in fig. 370, and then a thin casing of string board covers the whole. Fig. 369 shows detail of framing.

Handrails are often fixed to the balusters by means of an iron core, from $\frac{1}{8}$ in. to $\frac{1}{4}$ in. thick and of the width of the baluster, being screwed down to the latter and screwed up to the former. Treads and risers, in addition to being housed into the strings, should also have triangular glued blocks to stiffen them, on their under-side. The methods adopted for setting out handrailing for different staircases is a subject which must be dealt with by itself, as it would require several chapters to describe this somewhat intricate subject.

CHAPTER XXIII.

SHAPED WORK.

IN the different chapters already given the interest has lain rather in the construction than in the geometrical problems which they involve. In practice, however, the student will find that carpentry and joinery require a knowledge of a certain amount of descriptive geometry before designs or working drawings can be made of the different parts of a structure or piece of carpentry or joinery which is circular on plan or elevation, or in both.

Such features as *domes, niches, pendentives, angle-brackets, circular-headed sashes in circular walls, raking mouldings, elliptical lantern lights*, &c., require a considerable amount of practical geometry to draw them out so that the carpenter or joiner can execute them. As a matter of fact, this, of course, is usually done by the craftsman from the $\frac{1}{2}$-in. scale drawings supplied by the architect; and as no set of carpentry articles would be complete without some notices of the rules by which they are produced, it is proposed, therefore, to take an example of each of the above-mentioned features and work it out by means of an illustration.

1. DOMES are composed of a certain number of ribs placed vertically in planes, which in spherical domes would, if prolonged, pass through the vertical axis of the dome. In a true dome rising from a circular base it is evident that all the ribs have the same profile or contour; but in domes on polygonal plans the angle-ribs at the intersection of the sides of the solid are alone in planes which pass through the axis. In regard to the construction of ribbed domes, the ribs generally spring from a wall-plate or curb laid on the wall, and forming a ring strong enough to resist the lateral thrust of the ribs, the wall thus supporting only the downward thrust or weight of the dome. When the height of a dome is greater than the radius of its base, it is said to be surmounted; when less, surbased.

To take an easy example. Suppose we want to construct a flat or surbased dome over a rectangular room. The plan (fig. 371) shows the method of placing the ribs, springing from the base or wall-plate, and abutting on to the diagonal ribs, which meet in the centre of the plan. Fig. 372 is the section across the shortest diameter of the plan ; and fig. 373 the section taking on the longer axis. The curve of the rib passing across the centre on the longer axis and all the ribs parallel to it is found by dividing the segment (fig. 372) into a number of equal parts, and drawing lines from the division to meet the diagonal AB in the points $g, h, i, k,$ and from these points drawing lines parallel to CD, cutting EF, and produced indefinitely for the purpose of measuring all the heights of the ordinates on CD, a 1, a 2, a 3, &c., then by joining all these points the contour of the longest rib on the longer axis is obtained. It will

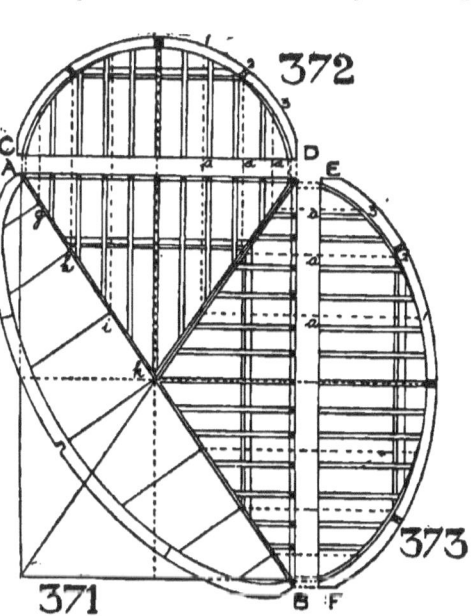

371. Plan of dome. 372 & 373. Sections of dome.

be seen that we start by setting out the ribs on a known segment of a circle, and by means of these heights transferred on the longer diameter we obtained the outline of the dome parallel to it. The curve of the diagonal ribs must also be obtained. This is effected by a similar operation by setting out equal spaces along the segment of the dome, producing these till they meet the diagonal, setting them off at right angles to it, and marking off the heights from the first figure. The purlins supported on the diagonal are produced from the fig. 372.

Figs. 374 and 375 show in section and plan a surbased dome on an octagonal plan. Fig. 376 shows the position of the ribs and the manner of finding the curve of the angle-ribs. Fig. 374 is a section on line AB.

The rib over MN is drawn in elevation at m 1 2 3 4 N and divided into equal parts from which ordinates are drawn to the chord line and produced to the line (on plan) of the angle-rib KL; from the points thus marked K w x y z L ordinates are drawn at right angles and, the height QN being transferred to them, give points which being joined form the curve of the angle-rib. In the same manner the curve of one rib of any shaped dome being given it is easy to find any of the others by means of ordinates set out at right angles to its base.

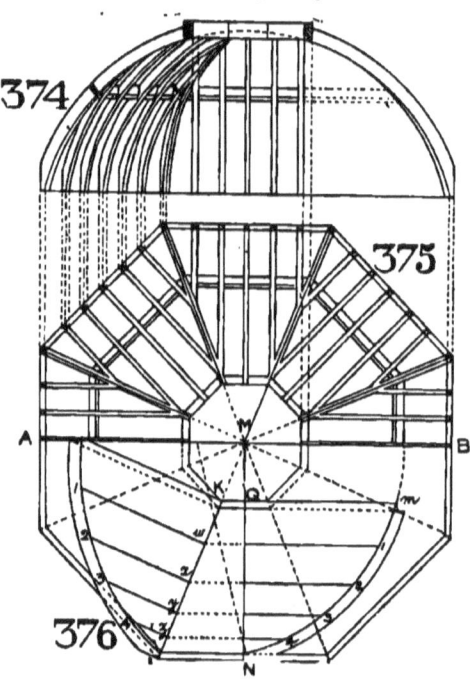

374. Section of octagonal dome.
375. Plan of octagonal dome.
376. Method of finding curves of ribs for octagonal dome.

2. NICHES.— The framework of these has often to be constructed in rough timbering, which is eventually covered with plaster, or in more finished work, when it is left exposed to view, as in the beautifully finished work of the Mediæval and Jacobean periods. At other times they are covered with boarding.

Figs. 377 and 378 show a spherical niche on a semi-circular plan, in section and plan. The ribs being all similar, the construction of this is precisely like that of a spherical dome. Fig. 378 is the plan showing the disposition of the ribs. Fig. 379 is a section showing the contour of one of the ribs. Fig. 377 is a section of the niche with the ribs set up by means of projection from the plan. The section (fig. 379) shows the bevelling of the back ribs, c, d, against the front rib at lmn on plan (fig. 378).

As another example of a spherical niche, fig. 380 shows such a niche on a segmental plan. Fig. 381 shows the plan and the method of setting out the ribs by drawing them to the centre, from which the plan of the niche is struck. Fig. 382 is the section through the upper part showing that the quadrant MN is drawn with the same radius as the plan of the niche, and the lengths of the back ribs are found by taking the distance cd, ef, from the plan, and setting them off on the line MP.

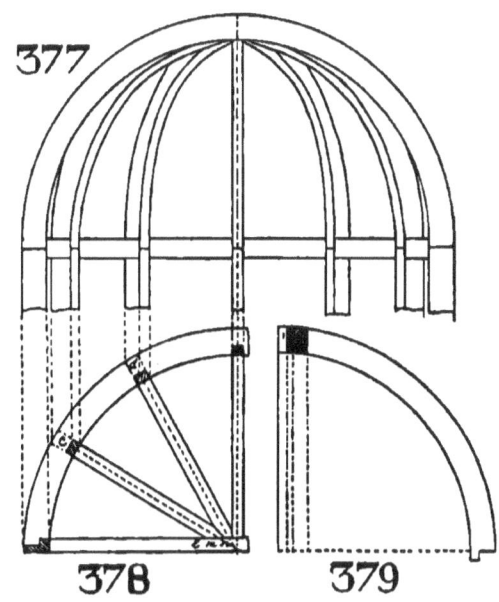

377. Section of niche.   378. Plan of niche.
379. Method of finding curves for rib of niche.

There are, of course, various other combinations of spherical, segmental, and elliptical niches, but enough has been shown to indicate the principle on which they are laid out. Let us take another type. Fig. 383 is the half-plan of an octagonal niche showing the ribs it is proposed to construct. Fig. 384 is the half-elevation of the niche. It being a regular octagon, the curve of the centre rib, PQ on plan, will be same as the half-front rib, NQ, shown in elevation. In fig. 385, therefore, draw the half-plan of the niche, NMOP, and draw the rib, PAB, equal to half the front rib shown in elevation in fig. 384. Divide the contour of this rib into any number of equal parts, and through these points of division draw lines parallel to the face of the niche, produced to meet the plan of the angle-rib, OB, in points a, b, c, d; on these points raise perpendiculars, and set upon them the heights, l_1, m_2, nB, &c. Join the points, and the resulting contour will be that of the angle-ribs of the octagon. The shaded part shows the bevels at the meeting of the ribs at the summit of the niche. We will give one other example, as follows:—The plan of a semicircular niche in a concave wall being given, to find the ribs. Let ABC, fig. 387, be the plan of the niche, AEC, showing the line of the wall in which it is situate. Join AC, bisect it at D, and draw the plan of the ribs from this point as a centre and the elevation as in fig. 386. The ribs have all the same curvature, being segments of a sphere, and the length of each can be obtained by describing the quadrant LMN, fig. 388, which is the true elevation of one of the

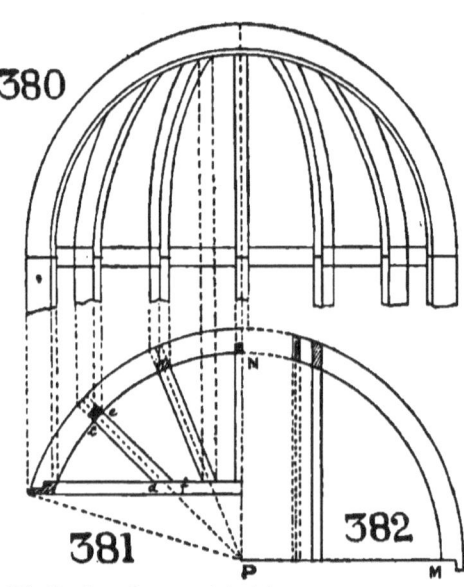

380. Section of segmental niche.
381. Plan of segmental niche.
382. Method of finding curve of ribs for segmental niche.

CARPENTRY AND JOINERY. 253

ribs if produced. The length of each intersection with the front rib is obtained by transferring the lengths from the plan to the radius in fig. 388, and drawing perpendicularly as shown. The face rib on the wall is a semi-ellipse formed as shown in figs. 389, 390, and 391. In fig. 390 make DC equal to DC in fig. 387, describe the segment CB, draw DB perpendicular to CD, and describe the curve of the wall EC. This is the plan of the niche, and in a similar manner fig. 389 is an elevation of the half-niche. In fig. 390 divide the curve EC into any number of equal parts 1, 2, 3, 4, 5, and draw 1*a*, 2*b*, 3*c*, 4*d*, 5*e* perpendicular to CD, transfer the lengths E5, 5 4, 4 3, 3 2, 2 1, 1C on this curve from D towards C on the line DC in fig. 391, and draw 1*a*, 2*b* perpendicular to DC. The heights 1*a*, 2*b*, &c., are formed by transferring *a,b,c,d,e* of the line DC, fig. 390, to the line AC, fig. 389, and make A*f* equal to DP, fig. 390. Then draw the perpendiculars *mg*, *nh*, &c., and transfer the heights to the corresponding ordinates in fig. 391, and to complete the curve more exactly divide the last space into *p*, fig. 390, and transfer to figs. 389 and 391 for the ordinates.

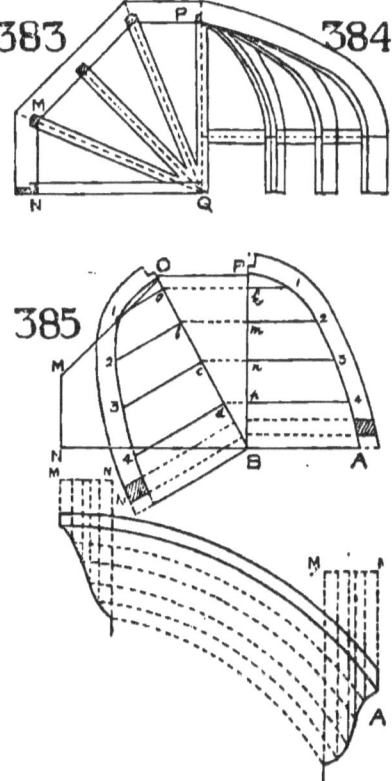

383. Plan of octagonal niche.
384. Section of octagonal niche.
385. Method of finding curves of ribs.

We have gone thus fully into dome and niches, as it is

the working out of these that the student will find himself most often called upon to undertake.

3. PENDENTIVES.—If a hemisphere or other portion of a sphere be intersected by cylindrical or cylindroidal arches, triangular spaces are formed which are called pendentives. The termination of these at the top will be a circle whereon may be placed a dome, which may be treated in a variety of ways. Take as an example the following problem:—

To cove the ceiling of a square room with spherical pendentives having a circular skylight in centre. This is a practical instance which an architect may have to draw out. Let ABCD, fig. 393, be the half-plan of the square room. From the centre describe the semicircle which is the plan of the hemispherical vault. On the side AC describe a semicircle representing the curve resulting from the intersection of the hemisphere by the plane of the side of the room. Then

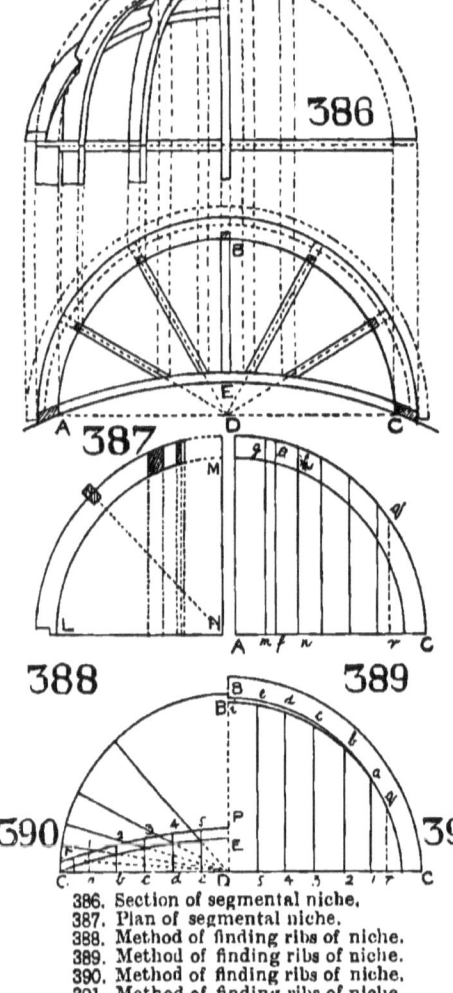

386. Section of segmental niche.
387. Plan of segmental niche.
388. Method of finding ribs of niche.
389. Method of finding ribs of niche.
390. Method of finding ribs of niche.
391. Method of finding ribs of niche.

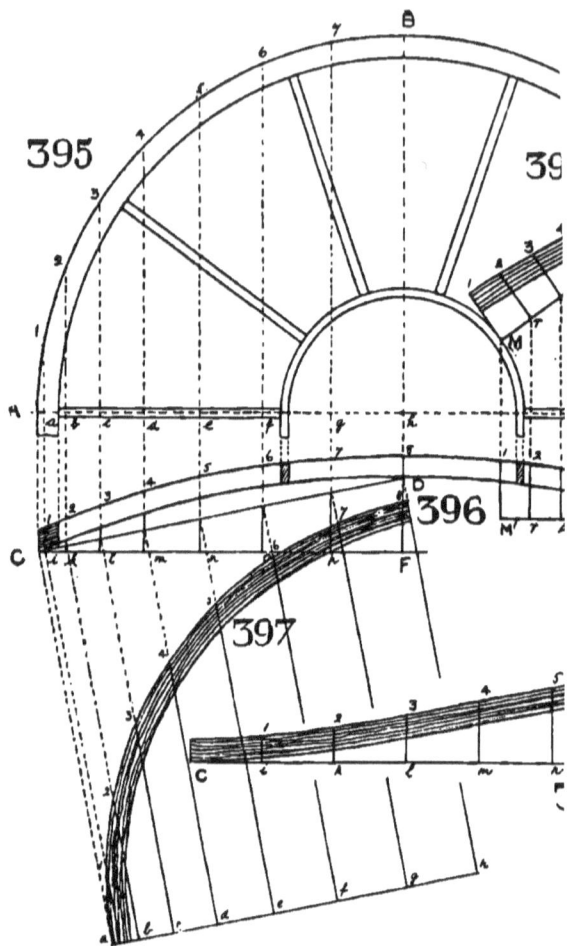

395. Section of circular-headed sash in circular wall. 396. Pla of rib for sash. 398. Method of obtaining mould to bend on up of obtaining mouldings. 401. Method of obtaining mouldings.

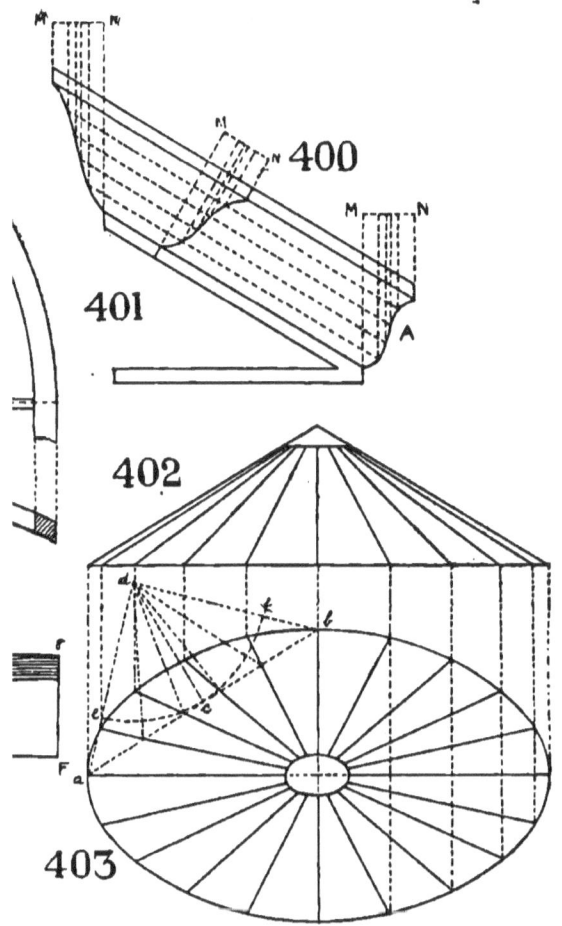

ed sash in circular wall. 397. Method of obtaining curve
d. 399. Method of obtaining radial bar. 400. Method
of skylight. 403. Plan of skylight.

CARPENTRY AND JOINERY. 257

set out the ribs on plan as it is desired to place them, first describing the circle for the skylight by two lines representing the thickness of its curb. The fig. 392 is a section taken across the centre of the plan. From O as a centre

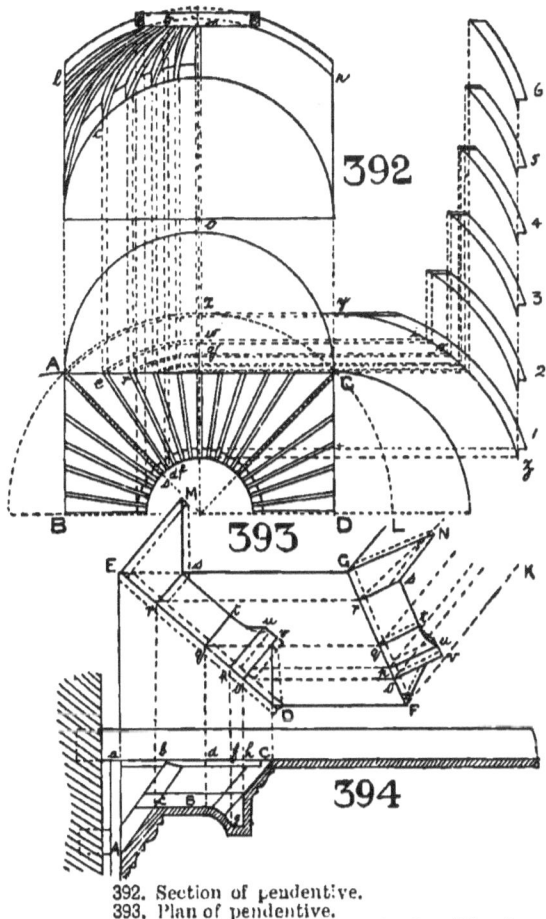

392. Section of pendentive.
393. Plan of pendentive.
394. Method of obtaining angle of wall bracket.

draw the semicircle representing the intersection of the hemisphere, and from the same point describe the segment *lmn*, representing the section of the spherical surface on the line BD. Project the curb from the plan

R

and find the intersection of the other ribs with the side and the curb by projecting lines from the plan from e to c and b to d. To find the true length of each rib proceed as follows :—In plan on the arc Ax draw xy parallel to the side of the plan; then with the same radius with D as a centre describe the arc yz, which will give the under-side of the rib As. From the centre describe the arcs ew and q and c, and draw wi, qv, to intersect yz in i, v, then iz and vz will be the true lengths of the ribs ed, rf.

By projecting the points i, v, z, and describing arcs with the same radius as yz, all the ribs may be drawn separately as shown in Nos. 1, 2, 3, 4, 5, 6. The bevel of the ends of the ribs is obtained by means of the double-dotted curves Ax, ew, rq.

4. ANGLE-BRACKETS (fig. 394).—The pieces of wood which support the laths of cornices, &c., are called brackets, and are finished to as near as possible the general outlines of the cornice to be supported. These may be either for internal or external cornices, forming re-entering or salient angles. Let ABC be the elevation of the cornice bracket, DE and FG the horizontal plan of the internal and external angle of the cornice.

From points which represent the change in contour of the bracket draw perpendicular lines, cutting DE in certain points.

Draw the line DF, FK, EG, GL, representing the plan of the bracketing and the parallel lines from the intersection $l m n o$, as shown. Then make EM and GN each equal to Aa, rs to bc, qt to de, pu to fg, and so on, and join the points so found to give the contour of the brackets required.

5. A CIRCULAR-HEADED SASH IN A CIRCULAR WALL is a problem of frequent occurrence, and the following is the method for setting it out. Fig. 395 represents the elevation and fig. 396 the plan of such a window. In order to find the lines necessary for forming the head for the sash divide the quarter circle AB into any number of equal parts; draw vertical lines intersecting the chord CD on the plan, fig. 396. From these intersections draw lines perpendicular to the chord CD, as shown in fig. 397. Draw any

line parallel to CD and set off on the perpendicular lines the heights $a1$, $b2$, &c., taken from the elevation. These points represent the upper curve of the sash-head. The under curve is found in the same way, and the lines being joined, the face mould for the head is obtained.

To find the mould to bend on the upper surface of the head proceed in this manner :—On any straight line CF, fig. 398, set along distances equal to those marked on the extrados to the sash; from these points raise perpendiculars equal to the distances $1i$, $2k$, $3l$, &c., taken from fig. 396.

Then a curved line drawn through these points as shown will be the development of one edge of the upper surface of the sash-head.

To find form of a radial bar (fig. 399). MN represents centre line of the bar. Divide MN into any number of equal parts, and from each of these let fall perpendicular lines intersecting the plan at M, N, as shown. From points r, s, &c., in line MN draw perpendicular lines M 1 and 2, &c., as shown, and mark off M. 1 and 2 equal to M 1 and 2 in fig. 396 and join points; this will give the radial bar.

6. RAKING MOULDINGS.—Mouldings which are not horizontal in the direction of their length are called raking mouldings. When a raking moulding has to intersect a horizontal moulding it is necessary, in order to produce a proper effect at the junction, that the correct form of section should be specially ascertained. In figs. 400 and 401, A represents the profile of the horizontal moulding. Divide this into any number of equal parts, and from these points draw parallel lines at the inclination of the raking moulding. From the points marked in the profile draw vertical lines as shown to any horizontal line as MN. Mark the widths on this line in a direction parallel to the rake, let fall perpendiculars in the manner shown, and the intersections with the lines parallel to the slope will give the form of the right section of the raking moulding.

7. A CONICAL SKYLIGHT ON AN ELLIPTICAL BASE, figs. 402 and 403, being planned, it is desired to divide the

curb into proportionate spaces. This is accomplished as shown by drawing the line *ab*, dropping a perpendicular from its centre, and making *cd* equal to *ca* and *cb*. Join *d a*, *b d*, and from *d* describe the arc *ecf*. Divide the arc into the same number of spaces that the quarter of the curb is desired to have. Through these points from the centre *d* draw lines to intersect the line *ab*. From the centre of the ellipse draw lines through these points to intersect the line of the curb, which will give the required points. We have thus traced a few of the problems of shaped work which the joiner has to carry out, and sufficient has been shown of the principles adopted to enable the student to apply them to any special case. The question of bevels will be treated separately in the next chapter.

CHAPTER XXIV.

BEVELS.

ANY angle that is not a right angle or an angle of 45 deg. (termed a mitre) is called a *bevel* angle. The exact angles which pieces of stuff make with adjoining pieces of stuff have to be accurately formed by the craftsman before they can be fitted together. The manner of finding the bevels (or *cuts* as they are called) as applied to different constructions is outlined in the following remarks, by which it will be seen that it is in general a simple matter, requiring only a sound knowledge of solid geometry.

Following the principle of tabulation adopted throughout this work as being most easy of comprehension, we may roughly class our remarks as under :—

1. Bevels for simple oblique work.
2. Bevels which occur in hip roofs.
3. Bevels which occur in valley roofs.
4. Bevels for purlins.
5. Bevels to splayed window linings.
6. Bevels to louvre frames, &c.

The instrument or tool called the *bevel* has been already described and illustrated in a previous chapter, fig. 29. It consists of a stock made of hard wood with brass mountings at each end and a blade held together by a screw-pin. The blade is made of a parallel plate of steel, with a slot allowing it to move up and down on the screw-pin at the end of the stock which passes through this slot. The slot in the stock permits the blade to be pivoted completely round on the screw-pin, and as it can be tightened in any position any angle can be made between it and the stock.

1. BEVEL FOR OBLIQUE WORK.—The simplest kind of bevel which the craftsman will have to perform may possibly be something like that shown in fig. 404, in which a vertical post is kept in position by a stay or strut which is inclined to the ground at an angle of 60 deg. As this strut has its side parallel to the plane on which the elevation is drawn, it will

be seen that the true length and correct bevels are shown in elevation, and are indicated by the thick lines enclosing the angle at which the strut has to be sawn at the top and the bottom. This is, perhaps, the simplest case it is possible to imagine, as the bevel can be at once set at 60 deg. for the lower angle and 30 deg. for the upper.

Fig. 405 shows a similar post, but in this case the strut has its sides at 45 deg. to the vertical plane. To construct the elevation draw the line mn to this inclination, and

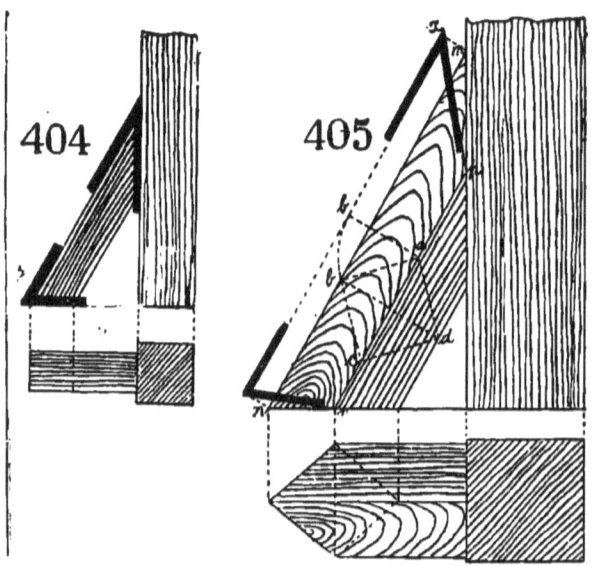

404. Simplest form of bevel. 405. Bevel between strut and post.

on this line draw a section of the strut to find the necessary angular points in elevation, which will be obtained by drawing lines through these points. The plan is projected from the elevation as indicated.

In order to obtain the bevels, the true length of the edges, being seen in elevation, it is necessary to develop the sides; this is effected by turning round the side ab, as shown, and drawing a line parallel to mn through this point, from m draw mx at right angles to mn, then draw from x to p, and the necessary bevel is obtained. In the same manner for the bottom bevel draw a line from *n* perpendicular to *mn*, and from the point where this cuts the developed edge draw a line from *q* which will give the required bevel.

The bevels for the under-side of the top and bottom junctions with the posts are obtained in a similar manner.

Fig. 406 shows a similar post and stay; but in this case the post has its sides making an angle with the vertical plane. Draw the line *mn*, showing the true inclination of

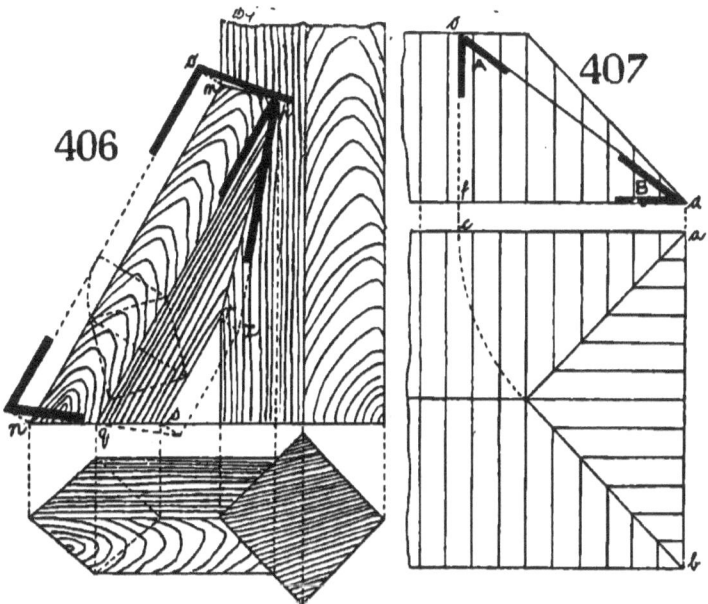

406. Bevel between strut and post, standing at an angle of 45 deg.
407. Plan and elevation of hipped roof.

one of the edges in elevation; on this line draw the plan as in the previous examples and draw the edges *pq* and *rs* parallel to the first edge; the point *p* in elevation is obtained by drawing a perpendicular from the point where it touches on plan; join *m* to *p* and *p* to *r*.

To find the bevels, develop the side and draw the line from *m* to *o* perpendicular to the edge *mn*, join *p* to *o*, which gives the bevel for the top side; the bevel for the under-side is obtained similarly by developing the under-side, and drawing a line from the point *r* to cut the developed edge in *x*; join *p* to *x*, which gives the necessary bevel; the lower edge is obtained in a similar manner.

In finding these bevels, the true length of the edge is obtained, and then the sides are developed; and where the true length of the edge is not shown, it must be projected upon a plane parallel to itself, thus showing its true length.

2. HIP ROOFS.—In its most simple form, the hip roof is a quadrilateral pyramid, each triangular side of which is a *hip*, and the rafter in each angle is a hip rafter. The *jack rafters* are those common rafters which, by abutting on the hip rafters, are therefore shorter than the length of the sloping side of the roof.

In a hip roof the following points are known:—The angle of the slope of the roof, the height of the roof, the length of the common rafters; and the points which require to be worked out are the angles which the hip rafters make with the wall-plate, the angles which the hip rafters make with the adjoining planes of the roof (called the backing of the hip), and the length of the hip rafter.

As a simple case, let fig. 407 represent a portion of a plan and elevation of a hipped roof, of which it is required to determine the length and inclination of the hip rafter. The pitch is assumed at 45 deg. for sides and end, and the angles *a* and *b* right angles. Take the length *ac* in plan and set it off from *d* to *f* in elevation, from the point *f* draw *fo* perpendicular (the point *o* being the height of the ridge), join *d* to *o*, then *do* is the true length and inclination of the hip, and the bevel at A shows the saw cut for the top end of the hip, in the same manner as B shows the bevel for the foot. The hip rafter is therefore seen to have a flatter inclination than the common rafters, *fo* being level with them at the ridge and at their feet; this is brought about by the hip having a longer distance to cover.

To take a more complicated example, in fig. 408 let ABCD be the plan of a roof in which it is required to find the length of the rafters, the backing of the hips, and the shoulders of the jack rafters and purlins. Draw GH parallel to the sides AD, BC dividing longitudinally the space to be covered, from the points ABCD with any radius describe the curves as shown, intersecting the sides of the plan, and from these points with any radius bisect the four

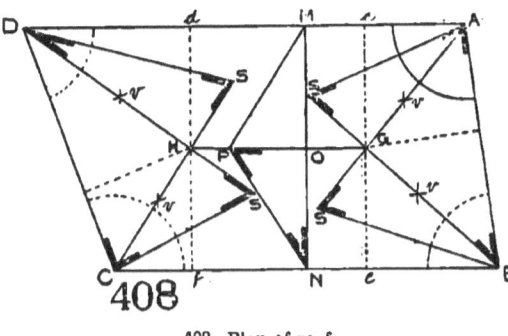

408. Plan of roof.

angles of the plan in the points v, and from ABCD draw the lines of the hip rafters through these points, cutting the ridge line in G and H. The dotted lines ce and df at right angles to the ridge indicate the plans of the end entire common rafters. From any line MN set off the height of the roof OP and join PM and PN, which show the true length of the rafter. From the points G and H, and at right angles to the plans of the hips, set off the heights GS and HS equal to the height of the ridge of the roof, then AS, BS, CS, DS are the true lengths of the hip rafters.

Take another case in which an irregular-shaped space has to be roofed in, and it is desired to keep the ridge level (fig. 409).

Let ABCD be the plan of such a roof, bisect the angles by the lines Ad, Bd, Ca, Da, and through a draw the lines am an parallel to the sides BC and AD, and meeting the lines bisecting these angles in m and n; the ridge wil then follow the line man, which will enclose a flat. On any

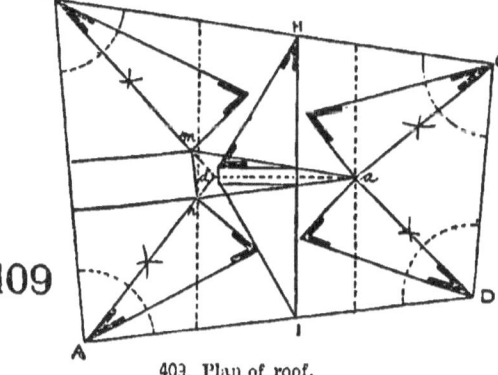

409. Plan of roof.

line HI set up the pitch of the roof as shown, and from the points *m, a, n,* and perpendicular to the plan of the hips, set off the height of the ridge, which, being joined to the angles of the roof, will give the true length of the hip rafters, and the bevels at their head and feet.

Another point which it is necessary for the craftsman to find out in setting up a hipped roof, is what is known as the *backing* or angle on the back of the hip rafter, formed by the meeting of the two planes of the roof.

In fig. 410 let AB, BC be the common rafters, AC the span of the roof, and DEF the plan of the hip. From E

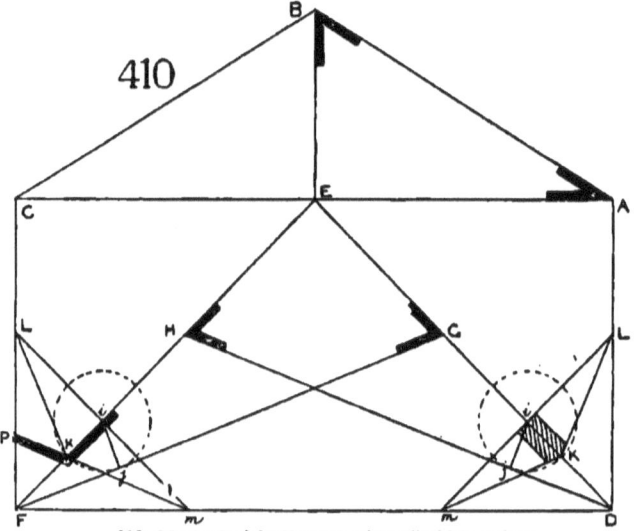

410. Method of finding "backing" of hip rafter.

set off the height of the roof EH, EG, equal to EB and join HD, GF, then HD, GF are the hip rafters. In order to find the bevel for the back of these, from any point *j* in FG draw the perpendicular *ji* cutting EF in *i*, and through *i* draw perpendicular to EF the line L*m* cutting FC and FD in L and *m*; make *i*K equal to *ij*, and join KL, K*m* and the angle EKP is the angle of the back of the hip rafter.

Take another case in which it is required to find the length and bevel of any jack rafter (fig. 411). Let ABC be the plan of one angle of a hip roof, BS the plan of the hip rafter, it is required to find the length and level of the jack rafter whose horizontal plan is DEF. At the point K where the longest side of the jack rafter meets the hip rafter, draw KM at right angles to side of jack rafter and make the angle KFM equal to the slope of the roof, draw also KP at right angles to side of hip KN and make KP equal KM. Produce KF and LE indefinitely and make KH equal to MF, then LK, GH is the length of the jack rafter. To obtain the bevel at the upper end draw NO at right angles to BC, making it equal to EG and join KO, then the angle OKH is the bevel of the jack rafter.

We have thus shown the principles upon which the bevels of the hip rafters and jack rafters in a hip roof are found, and more difficult problems are worked out on the same lines.

3. VALLEY ROOFS.—A valley is the internal meeting of the two inclined sides of a roof. The rafter which supports the valley is called the valley rafter.

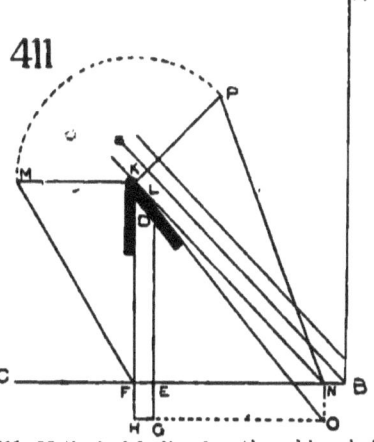

411. Method of finding length and bevel of jack rafter.

Let us take an example. Let fig. 412 represent the half-plan of two roofs of different pitches meeting in a valley. Two jack rafters are shown, one to each roof. The larger roof is inclined at 30 deg., and the pitch of the small one is found by making the total rise the same as the large one.

To find the bevel for the short jack rafter, raise perpendiculars from 3 and 4 in the plan until they cut the slope of the roof (placed temporarily at N). After making the width equal to that shown on plan, take a perpendicular across from 4, meeting the opposite edge in 5 ; join points 3 and 5, and the edge-cut is complete. The edge-cut for

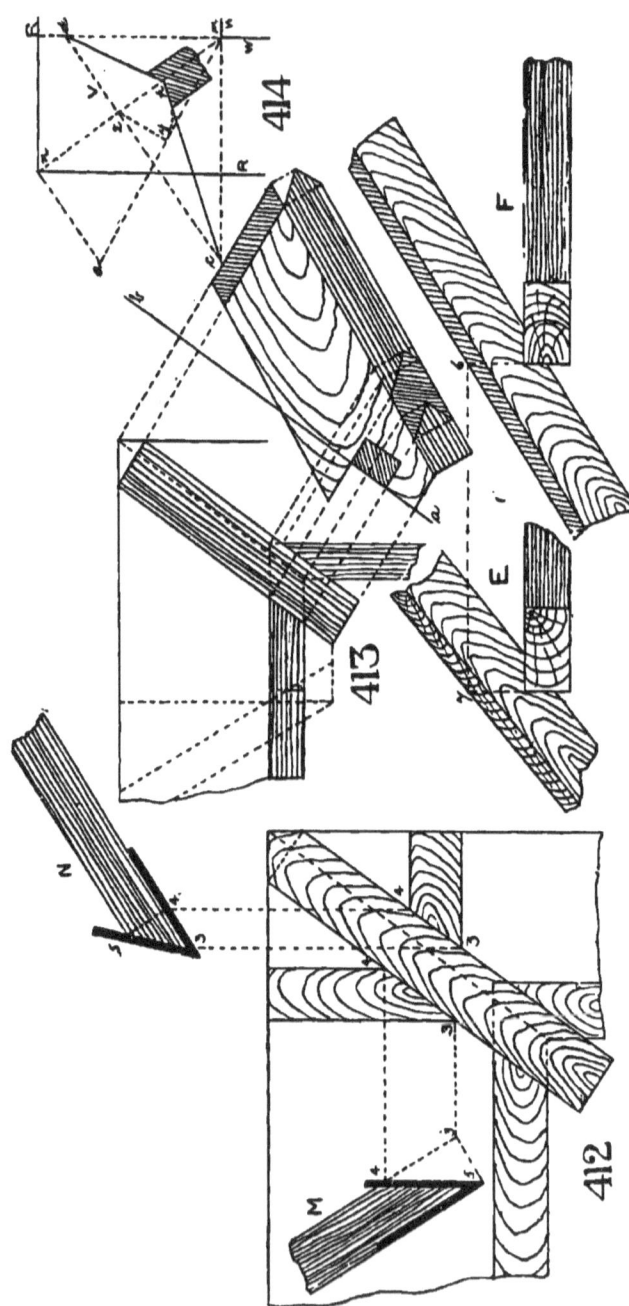

412. Half-plans of two roofs of different pitches. 413. Plans of two roofs of different pitches. 414. Method of finding "backing" on upper side of valley rafter.

the rafter to the wider span is obtained in a similar manner (see M). In constructing this valley, the jack rafters are kept below the edge of the valley to allow for the thickness of the boarding. In the figures E and F this is indicated. The figure at F represents the roof of 30 deg. span with jack rafter, wall-plate, and valley board. Then, in order to find the necessary notch in the rafter of the quicker-pitched roof, draw the perpendicular to 6 ; carry this point by means of the dotted line to 7, which will give the position within which the boarding must be kept at this point.

In order to find the bevels of the valley rafter we must proceed as follows :—The fig. 413 shows the same valley rafter as last described, the ridges and wall-plate being also marked ; on the line *ab*, parallel to the valley, is set up the true pitch and length of the valley. The student will notice the notching on the wall-plate, and the bevel at the top end, where the valley comes against the ridge. To find the bevels develop the edge, and project each point, as shown both in the lower and upper ends. We have still to show the backing or bevel on the upper side of the valley rafter. Fig. 414 shows the method of finding this. W, R, and V represent the wall-plates, ridge, and valley rafters of the roof ; at any point on the plan of the valley draw *cd* at right angles to *mn*, meeting the line of the wall-plates produced in *c* and *d*; from the point *n* draw *na* at right angles to *nm*, and equal to the rise of the roof. By joining *a* to *m* the length and inclination of the valley is shown ; from *x* draw *xy* perpendicular to *am*, and set off from *x* the distance *xp* equal to *xy*, join *cp* and *dp*, and the true shape of the angle of the valley rafter is obtained.

4. BEVELS FOR PURLINS.—To find the bevels for purlins passing round a hipped roof proceed as follows (fig. 415) :— Let E be the purlin, BC the slope of the roof, and AD the plan of the hip. From F describe any arc HN and draw HJ and NO perpendicular to the diameter HN. From I and L, where the upper sides of the purlin cut the circle, draw IK, LM, parallel to HJ, and FG, also parallel to HJ. From M and K draw MO, JK, parallel to BD, and join GJ and GO. Then FGO is the down bevel of the purlin, and FGJ its side bevel.

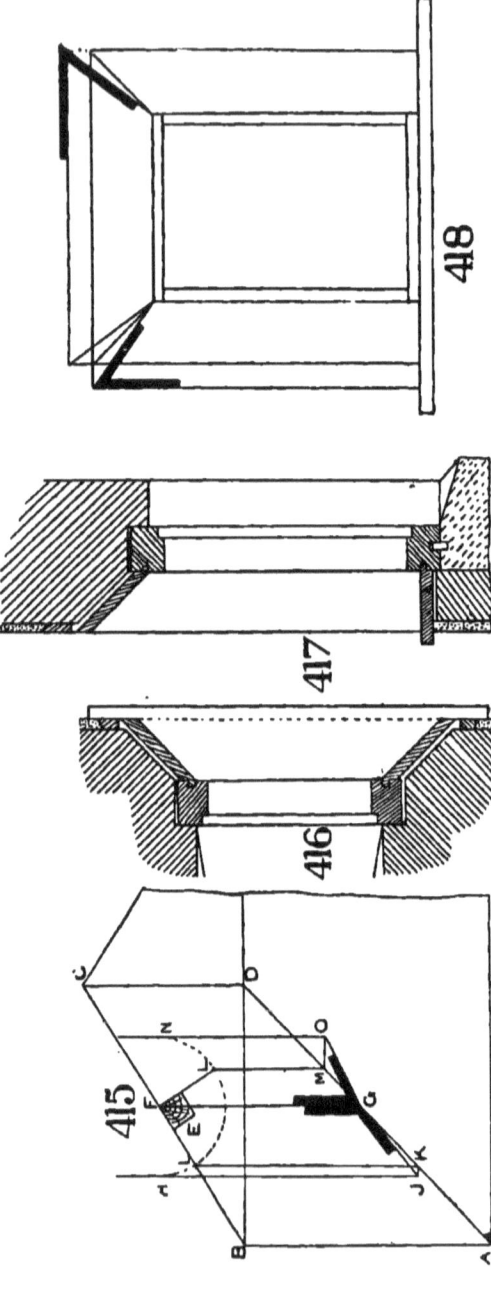

415. Method of finding bevels for purlins in hipped roof. 416. Plan of splayed window. 417. Section of splayed window. 418. Elevation of splayed window.

5. BEVELS FOR SPLAYED WINDOW LININGS.—Linings to windows are often splayed to admit more light, not only at

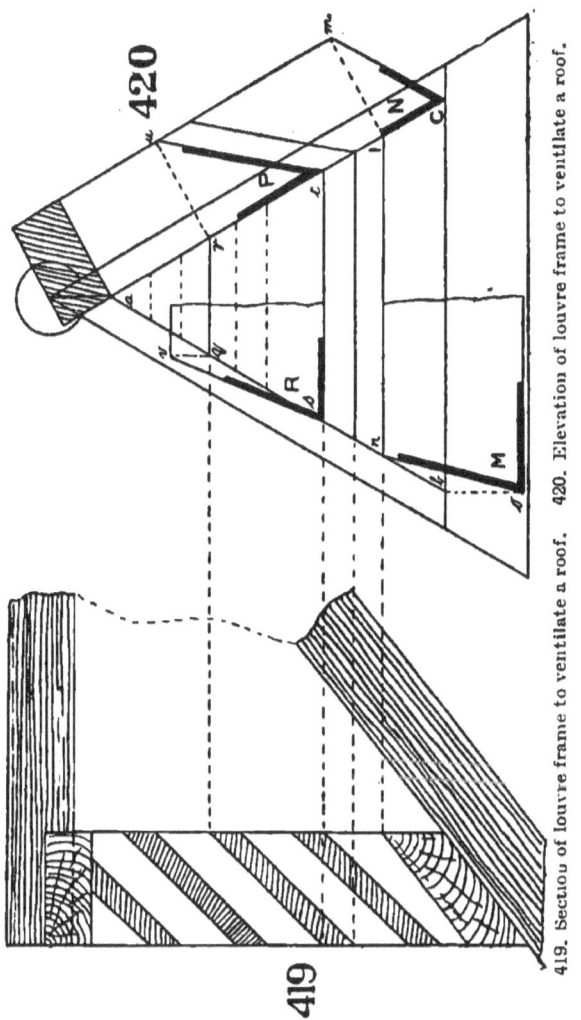

419. Section of louvre frame to ventilate a roof. 420. Elevation of louvre frame to ventilate a roof.

the sides, but also at the top. Figs. 416, 417, and 418 show such a window in plan, section, and elevation. The bevels or cuts for the junction of the upper or side linings, shown in black lines, hardly want further explanation. It is evident that the proper bevel can only be obtained by developing the side and head respectively, and thus obtaining the true cuts.

6. BEVELS FOR LOUVRE FRAMES, &C.—Figs. 419 and 420 are examples of a louvre frame for a ventilator in a roof, of which it is required to find the bevels of the various parts.

Fig. 420 shows the elevation and fig. 419 the section of the ventilator. In the elevation *abc* represents the inside of the sloping frame. In order to determine the true shape of the sides they should be developed as shown (*i.e.*, the real width drawn) and from the point *l*, where the side meets the back of the sill, draw a perpendicular meeting the inside at the point *m*; then by joining *c* to *m* we shall obtain the bevel N, which will enable the sides to fit properly on to the top of the sill. Again, the sides being grooved into the sill it is necessary to find the bevel from this groove; for this purpose develop the top side of sill, as shown; from the point *b* draw a perpendicular to meet the developed edge, join *no*, and the bevel M for the groove is obtained. Next, as to the louvre-boards themselves, we must determine the bevel for the groove in the sides into which these are fixed, and also the bevel to which they are cut. As to the former, the louvre *qrst* is shown in elevation, and from the point *r*, where the louvre cuts the back edge (see section), draw a perpendicular meeting the edge of the side developed in *u*, join *u* to *t*, and the bevel for the groove is obtained, marked P.

To find the side-cut of the louvre, develop, as always, the top face of the board, and from the point *q*, where the top edge cuts the side of the frame, draw a perpendicular meeting the developed edge in *v*; then, by joining *s* to this point, the bevel R is obtained for this louvre-board. The others follow in a similar manner.

We have now glanced at a few of the principal bevels or cuts used in carpentry and joinery, and if, as we hope, the principles of forming these have been grasped, they will be easily applied to any case which will engage the craftman's attention.

CHAPTER XXV.

RODS.

BEFORE any work of importance can be carried out it is necessary that at least some part of it should be drawn out full size so that it can be referred to at any moment. This is done on a board which is called a *rod*. The old method of setting out rods was to chalk an ordinary deal board and to mark thereon with pencil lines, but this is found not to be a good method, owing to the chalk wearing off so easily and thus obliterating these lines.

The method now usually pursued in good workshops is to mix up a bucketful of whiting to the consistency of cream and to add a cupful of glue. The pail is then placed on the stove when the whole is well stirred and mixed together. There should be just sufficient glue to "fix" the whiting, in order to prevent it rubbing off too easily, but there should not be sufficient to make it string, as it then is difficult to draw upon. A pine board, 11 in. wide by $\frac{3}{8}$ in. or $\frac{1}{2}$ in. thick, is generally employed, and after it has had a good coat of the whiting, made as described, it is left three or four hours to dry, and is then papered down with glass-paper, when it is ready for setting out.

It is well to remember that in setting out rods, excepting only in the cases of very intricate or circular work, sections (either vertical or horizontal) are only employed.

In the cases of setting out rods for windows and doors, it is most important to work from the *brick opening*, as very few bricklayers are always particular to follow out the architect's drawings very exactly, and, in fact, it is not always possible to do so, as the bricklayer is limited to the size of his bricks and the courses of the brickwork, and also many bricks vary considerably in size.

First, therefore, obtain the size of the opening, the width of which in the case of a window is generally between the brick reveals, the pulley styles lining up from these. The height being from the stone sill to the underside of the gauged arch when the latter is horizontal; when it is segmental or circular, of course the height is taken to the springing, and the rise is then added. It is almost a universal practice to treat all openings as though square in the first instance, and then to cut a template in the case of any arch not either of the ordinary circular or segmental type. Perhaps the method of setting out rods will be best explained by taking a simple example. Figs. 421—423 represent the plan, section, and part elevation of a 2-in. framed partition which is purposely made of a heterogeneous design to illustrate better the method of setting out of fig. 424. Let CDEF represent the rod about to be set

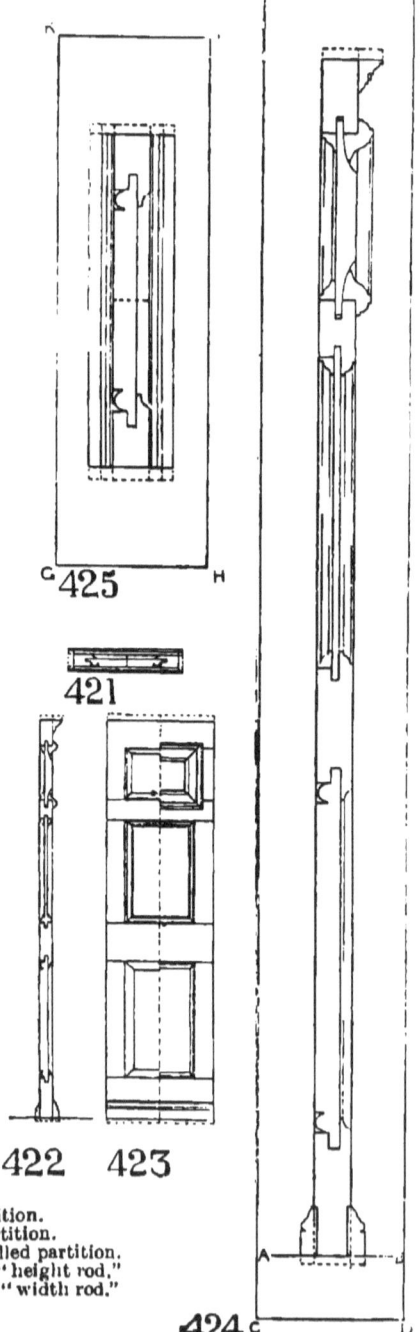

421. Plan of panelled partition.
422. Section of panelled partition.
423. Half-elevations of panelled partition.
424. Method of setting out "height rod."
425. Method of setting out "width rod."

out, and from which the joiner will take his measurements.

Commence by drawing a floor line AB on the rod. In laying down the bottom rail on the rod, a little extra depth, shown by dotted lines (say an inch for soft woods and about half an inch for hard woods), is allowed for scribing to the floor in case of any irregularity or unevenness of the latter as laid. Mark off the total height of the partition, which in this case is regulated by the ceiling-level, and again make an allowance for the purpose previously described for the bottom rail; next mark off the top edge of the middle rail, and set down therefrom the actual depth of the rail. It is usual to set off all the top edges of the rails from the floor line, and then to mark down therefrom their respective widths. Then mark off the middle rail $\frac{1}{2}$ in. within its lower edge for the panel groove (see fig. 422). Mark on the height of the frieze rail from the floor line, and set down therefrom the depth.

Next set down from the ceiling line the depth of the top rail. Now proceed to line in the panels, commencing from the bottom. The thickness of moulding to be used must be ascertained. In this case we have a $\frac{3}{4}$ in. moulding to deal with, so draw a vertical line $\frac{3}{4}$ in. from the *face* side (the *face* side of the work is the principal side, the other side being termed the inside), this will leave in thickness a $1\frac{1}{4}$ in. bead flush panel. Draw the section of the beads at the top and bottom of the panels, and allow a $\frac{1}{2}$ in. tongue at either extremity. The middle panel being glazed with $\frac{3}{8}$ in. plate glass, draw two vertical lines $\frac{3}{5}$ in. apart, equidistant from the centre of the thickness, and on either side draw the section of the moulding at the top and bottom. This moulding would be stuck on the framing on the face side, *i.e.*, worked in the solid; on the inside this moulding would be screwed on with cups and brass screws. The top panel is raised; this must be drawn to the section (fig 422) showing a $\frac{1}{2}$ in. flat surface of sufficient depth top and bottom to take the bolection-moulding which overlaps the framing $\frac{3}{8}$ in. To fix this moulding the joiner would screw it from the inside of the panel; these screws in turn being covered by the inside moulding which would be glued or dowelled according to the quality of the work. Draw the moulding of the panel and the *fielding*, *i.e.*, the flat surface or "field" of the panel, and then draw in the moulding on the inside.

Now draw in the skirting on the face and inside, making a sinking of ⅛ in. within the thickness of the lower rail so as to prevent the joint showing in case of shrinkage, and allow for scribing to floor; then draw on the cornice at the top on the face side, or allowing in the case of a thin one a similar sinking as before described for the skirting.

The rod is now complete as regards the vertical section which is called by the joiner the "height." The "width" or horizontal section is set out as follows:—In fig. 425 let GHIK represent the width rod, commence by drawing in the outside or "clear" width (in this case 3 ft.), allowing outside this ½ in. extra width for scribing (see dotted lines), mark on this the width of the style on either side. Draw two lines 2 in. apart to represent the thickness of the framing and also two lines 1¼ in. outside these to indicate the thickness of the skirting, and outside this the line representing the greatest projection of the cornice.

Now draw in the panel, which can be gathered from the "height" rod, showing here the section of the ⅜ in. bead as this is "bead and flush."

The "height" and "width" rods are now complete for this simple piece of panelling.

In more elaborate works such as a blank screen with mouldings, breaking round pilasters, all these breaks would have to be carefully drawn in for the purpose of obtaining mitres as the joiner measures these direct off the rod.

The methods above described are usual in the case of all joiner's work with the addition, in the case of shaped work, of a full size elevation, which is made on a rod composed of boards, usually glued together and provided with ledges on the back to strengthen it.

CHAPTER XXVI.

WORKSHOP PRACTICE.

It is proposed to treat briefly on the above subject under the following heads :—
1. Framing up.
2. Fastenings.
3. The use and sharpening of tools.
4. The care of tools.

1. FRAMING UP.—Great care should be taken in framing up work, as the stability of framing depends so much on the close and proper fitting of the joints. The following is the method usually adopted in putting a door together :— All the styles and rails, &c., should first be fitted, and every part marked with a distinguishing letter or figure, so that the proper position of each piece may be clear. Commence by putting the lower rail in the bench-screw, and drive the munting into its mortise, next drive on the middle rail and put in the lower panels, then lower this part of the framing on to the floor and drive the top munting into its mortise. Next put in the top panels and drive on the top rail, then turn the framing on the ends of its tenons on one of its sides and drive on the style. Turn the door over and drive on the other style, by this means all the parts are put together without any bruising of the shoulders.

This having been accomplished, lay the door on the bench upon two pieces of stuff, say 3 in. by 3 in., "out of winding," as this ensures the door itself remaining true, and then "try up" with the cramps. Then knock to pieces, and well glue both mortise and tenons, quickly drive rails on the muntings, then place mortises of styles on ends of tenons of rails, gluing both sides of tenons and mortises, and drive home and well cramp up, using two cramps, one at each end of the door. Then dip the wedges in glue and drive them home; after which leave the door till the glue is thoroughly dry.

The craftsman should be careful in the selection of his glue; that of a pale shade is generally superior to the darker shades. It should be prepared by being broken up in small pieces, and left to soak in water that just covers for about twelves hours; it may then be put in the glue-pot and thinned by hot water as desired. Work to be glued should be thoroughly dry, and be made perfectly clean and smooth, and the glue be applied as hot as possible. Glued articles should not be exposed to a cold temperature till the glue is set, and this it will not do in a freezing atmosphere. When glued work has to be exposed to the influences of the weather, linseed-oil and white lead are sometimes mixed up with the glue.

2. FASTENINGS.—*Bolts* are generally used for the purpose of giving added security to joints, though they weaken the wood through which they pass by severing the fibres, and are liable to become loose through the shrinkage of the timber; but this is generally obviated by one end having a solid head and the other a screw, on which is a movable nut, which may be tightened up. The following proportions for nuts and bolts are generally used:—

Diameter of head and nut, $\frac{1}{4}$ of diameter of bolt.

Thickness of head, $\frac{3}{4}$ of diameter of bolt; depth of nut, the same as the diameter of the bolt.

Washers should be about three times the diameter of the bolt and half the thickness of the head.

Drawboring consists of forcing the shoulder of a tenon close up to the mortise, and is accomplished by making the hole in the tenon nearer the shoulder, instead of exactly coinciding with the holes in the cheeks of the mortise. An iron pin is then forced through, the latter being afterwards withdrawn and an oak pin inserted.

Nails are generally used for securing boards to beams, and are much in request for temporary purposes and in inferior work. They are generally known by their weight per thousand and by their length in inches. Formerly they were known by their price per hundred, *e.g.*, "tenpenry" nails meant tenpence per hundred.

Pinning consists in driving an iron or hard wood pin through the different pieces of wood, forming a joint to hold them together. Care should be taken that the pin be not placed too near the end of a tenon, or the latter may shear.

Shoes of iron are sometimes used to receive the ends of the principal rafters, in which case a tie-rod instead of a tie beam is generally used.

Socket pieces are occasionally used to receive the upper ends of the principal rafter, as shown at S in fig. 60.

Spikes are large wrought nails used for heavy work, as in bridges, and often in large temporary erections. The larger ones generally have square, flat heads.

Screws used by the carpenter and joiner are called *wood* screws, to distinguish them from screws used in metal work. The thickness of screws is identified by numbers from oo to 40, the former being about $\frac{1}{32}$ in. in diameter and the latter $\frac{3}{8}$ in. Thus each number varies about $\frac{1}{84}$ in. from that number immediately above or below it. They are generally classed according to the shape of their heads, and are sold by the dozen or gross. There are many special forms of screws, such as handrail screws, dowel screws, coach screws, &c., which space prevents describing.

Straps are sometimes used instead of bolts to strengthen joints, and they have the advantage of not weakening the fibres.

It is desirable that they should be fixed as much as possible so that the strain should be in the direction of their length.

Stirrups are straps that support timbers, as shown at B in fig. 145, and the *gib* and *cottar* shown in fig. 61 is a form of strap supporting the tie-beam.

Trenails are pieces of hard wood used instead of iron nails for forming strong joints, and are useful when the work has to be exposed to the weather, as oxidisation is thus prevented. They should be split from the log so that the fibres are not injured.

3. THE USE AND SHARPENING OF TOOLS.—*Saws* are generally the first instrument used by the craftsman, so that it will probably be well to treat of these tools first. In

using the ripping-saw, the board or plank to be cut up should be laid between two sawing-stools, and the right knee should be firmly placed upon the "stuff" so as to keep it steady. The saw should be grasped by the right hand, with the index finger extended against the right side of the handle. The saw should be worked up and down in the same plane through the pencil line which has previously been marked on the stuff. It is well to constantly observe both sides of the saw to see that no divergence is taking place. As the work proceeds it is well to drive a wedge into the slit made, as this keeps apart the two pieces to be severed, and allows a freer passage for the saw.

After the sawing has been commenced by a few short and gentle strokes, the length and strength of which should be gradually increased, the whole length of the saw should be called into use, care being taken not to draw it out of the cut, as it is likely to be injured by striking against the wood in the return thrust. Great care should be taken, especially when using both hands, not to jamb the teeth into the wood, but to work the saw evenly and smoothly, as otherwise the saw may be rendered useless by being "crippled."

A slight deviation from the straight line may be rectified by twisting the blade, but this renders the saw liable to bind in the stuff.

When cutting through wood more than about $1\frac{1}{2}$ in. thick it is usual to lubricate the saw with oil. If the stuff is more than 2 in. thick it is a good plan to mark on both sides the desired position of the cut, and to occasionally turn it over so as to cut from the opposite face, as this conduces to straighter and more regular sawing. When using a saw care must be taken to cut just a little thicker than the actual width required, owing to the thickness removed by the "set" of the saw. This especially applies to back saws when cutting tenons, &c., and also where wood is not going to be planed up afterwards.

A saw is sharpened by three different operations, which are performed in the following order :—viz., *filing, setting*, and *gumming*. After the saw has been secured in a clamp, teeth uppermost, the teeth are dressed with the file at the desired angle. The saw is then secured in a horizontal vice, and the alternate teeth are *set*, *i.e.*, struck in a uniform manner with a hammer designed for the purpose, so as to bend every tooth at the same angle from the true horizontal. The saw is then reversed and the same operation is performed on the alternative teeth on the opposite side and in an opposite direction. *Gumming* is the process by which the throat of the teeth is deepened, and this is effected by means of punches. The three operations described are, at the present day, usually performed by machinery where possible, owing to the saving of labour and to the greater exactness obtainable by this means.

Planes.—The first thing is to adjust the irons. These should be taken out of the plane by grasping below the fork of the wedge with the left thumb and placing the fingers of the left hand round the plane and planting the back on the bench : a few smart taps on the nose of the plane will loosen the wedge, and the irons may be taken out.

The screw connecting the cutting and back irons may now be loosened, and the former, if necessary, sharpened.

When this is accomplished, as described later on, screw up the irons tightly together, the cutting-iron being about $\frac{1}{32}$ of an inch in advance of the back-iron. The irons are then placed in the plane, which is held at an angle towards the craftsman with its back end on the bench, so that he may look along the sole of the plane.

When the edge of the iron projects the required distance beyond the sole of the plane, the iron itself is kept in position by the left thumb, and the wedge is pushed into its place by the right hand, and is gently tapped " home."

The set of the plane may be adjusted during use by tapping the iron or the nose as occasion may require, and when the iron is not set squarely with a sole a side tap is generally sufficient to correct it. The wood to be planed should be laid quite flat upon the bench, and pressed against the bench-stop to prevent it from shifting. The plane must always be used in the direction of the grain, and if the latter runs in different directions, it must be turned about accordingly. When in use, the handle of the *jack-plane* should be grasped by the right hand with the forefinger extended against the wedge. The left hand should hold the fore part of the plane, with the thumb down the edge nearest the operator.

The *trying-plane* should be held in a similar manner, but the exertion to use it differs in this respect, that whereas the pressure of both arms should be uniform throughout in the case of the jack-plane, in the case of the trying-plane for the first half of the stroke the left arm exerts the most pressure, but for the latter half the right arm does most of the work. For "shooting" work the trying-plane is held with the fingers under the sole, which thus act as a kind of gauge for keeping the plane on the edge of the stuff.

The *smoothing-plane* is held by grasping it behind the cutting-iron, the fingers and thumb of the left hand being in front and pressing it down on the wood. The sole of all wood planes is, of course, liable to wear away, and must occasionally be planed up true. The liability to wear away is reduced by occasionally well rubbing the sole with oil. When a plane-iron is required to be sharpened it must be examined, and if the edge is found to be only "dull" it can be sharpened on an oilstone in the following manner:—Hold the iron with the right hand towards the top with the fingers on the under-side, and place the fingers of the left hand together and across the upper surface of the iron. The iron is then held at an angle of about 40 deg. and rubbed carefully on the oiled stone for its whole length, as it is most desirable to wear away the stone as little and as truly as possible. It is also important that the same angle should be observed throughout the stroke, or the cutting-edge will be convex instead of quite flat.

A blunt iron can generally be detected by a whitish, worn appearance, whereas a keen edge is invisible. The "wire" edge on the back of the iron should be removed by placing the back quite flat on the stone, and gently rubbing backwards and forwards for a few strokes. When the original bevelled edge made by the grindstone, which is generally at an angle of about 25 deg., has been much encroached upon by frequent sharpening on the oilstone, the grindstone must be again requisitioned, and a similar bevel to the original one must be ground on the iron. The latter is generally placed in a "support," which keeps it at the same angle, and thus a true bevel is obtained. The grindstone should be turned towards the operator, and should always have a trough underneath, so that when revolving it may be kept wet.

Chisels receive guidance only from the hands of the operator, and so require especial care and skill in using. *Paring-chisels* are used without the aid of a mallet, the left hand holding the wood, and being always kept behind the edge of the tool, so that it may not be injured in the event of the chisel slipping. In paring along the grain, the index finger should be extended along the tang of the tool, but when working against the grain the handle should be grasped by all the fingers.

The *mortise-chisel* is used in the left hand, while the right manipulates the mallet. Great care must be taken to make the edges cut by the chisel perfectly square. It will be noticed that if the flat side of the chisel be held against the shoulder to be cut away, the chisel must draw in and cause the joint to be hollow, whereas if the bevel side be held against the shoulder the contrary effect would be the result. Consequently, the latter method should be adopted, and the superfluous material can be carefully pared away at the finish with the flat side against the shoulder.

Always commence a mortise by cutting wedge-shaped chips from the centre, working alternately from each side.

The remarks made as to sharpening planes apply equally well to the sharpening of chisels.

The *gouge* is used in the same way as the paring-chisel, and does not require further description. *Scribing-gouges* and bent tools are sharpened by being rubbed with a small piece of oiled Turkey stone.

Gauges should be used in the left hand when adjusting and using, as this saves much time by obviating the necessity of the craftsman to alter his position at the bench.

Grindstones consist of a wheel of sandstone mounted on axles revolved by means of a handle. They are too often neglected, because, as a rule, it is not the duty of any one in particular to look after them. They should not be exposed to the rays of the sun, as they become so hard as to be worthless, and the water in the trough should be drawn off after every operation, as if left to stand in water the grindstone becomes soft in that part.

Oilstones are of several kinds, the "Washita," perhaps, being the best value for the price. A case should always be made for them, and they should be wiped after use, with a handful of shavings. Salad-oil only should be used, and cleansing with paraffin is not to be recommended, as it hardens the stone and thus reduces its cutting power.

4. THE CARE OF TOOLS.—The craftsman should always take a pride in having his tools well kept. The wooden parts of tools, such as the handles of chisels and the stocks of planes, are often soaked in linseed-oil for a week when fresh from the makers, and then rubbed with a soft cloth for a short time every day for a fortnight. This treatment preserves the wood, and at the same time gives a good surface. All steel articles can be preserved from rust by placing a lump of freshly-burnt lime in the tool-chest. To prevent the lime from soiling the cabinet it should be placed in a muslin bag. To preserve a polished surface the metal may be covered with a mixture of one part of resin to six of lard, dissolved slowly together, and then thinned with a little kerosine. It should be very lightly rubbed on, and can be easily removed when required. Rust may be removed from metal by rubbing well with sweet-oil, and then leaving it to soak for two days in the oil, after which it should be rubbed with finely-pulverised, unslaked lime. Another method is to immerse the article a short time in a solution of $\frac{1}{2}$ oz. of potassium cyanide to 4 oz. of water; it should then be taken out and well brushed with a cream paste composed of Castile soap, whiting, potassium cyanide, and water.

INDEX.

A

	PAGE
Acacia wood	13
African oak	10
Alder wood	13
American oak	10
,, plane	13
,, yellow pine	9
Angle-brackets	258
Aprons	225
Arch, Trimmer	132
Architraves	215
Arrises, timber	149
Ash, its use, &c.	12
Astragal moulding	176
Australian pine "Kauri"	2
Austrian oak	10

B

	PAGE
Backings	208
Balk of timber	21
Balusters	237
Bar, Sash	202
,, ,, vertical	202
,, ,, horizontal	202
Barrel bolts	200
Batten	21
Bead	176
,, inside	203
,, outside	203
,, quirked	168
,, double and quirk	176
,, and double quirk	176
Beam, Tie	48
,, Hammer	58

	PAGE
Beam, Dragon	55
,, Flitched	114
,, Curved	116
,, Indented	116
,, Joggled	115
,, Coke-breeze	206
,, Formula for breaking weight of	114
,, Formula for calculating strength of	115
Beech, its use	13
Bench, Carpenter's	22
Bevels	261
,, for hip roofs	264
,, ,, louvre frames	272
,, ,, purlins	269
,, ,, splayed window linings	271
,, ,, valley roofs	264
,, ,, oblique work	261
Binders	123
Birch wood	13
Blinds	209
Board, Window	208
,, V-jointed	166
Boards, Floor	135
,, ,, folded	135
,, ,, butt-jointed	137
,, ,, rebated	137
,, ,, rebated, grooved and tongued	138
,, ,, ploughed and tongued	138
,, ,, grooved and tongued	138
,, ,, rebated and filleted	138

	PAGE
Boards, Floor, dowelled	139
,, ,, for special purposes, dancing, &c	140
Bolection moulding	188
Bolt, Espagnolette	212
Bowtell pointed moulding	181
,, round ,,	180
Boxed frame	203
,, gutters	57
Brace and bits	27
Braces for scaffolding	101
Brackets, Angle	258
Bressumers	150
Bricks, Wood	206
Bridge	63-75
,, Trussed girder	68
,, Timber suspension	70
,, Close-ribbed	72
,, Flat lintel	66
,, Braced lintel	67
,, Trestle	64
,, Closed rib	72
,, Curved rib	70
,, Bamberg	73
,, Trajan's	63
,, Cæsar's	63
,, table of least rise for different spans	65
,, formula for finding curvature	71
,, average strain per foot super.	75
Bridle joint	35 & 37
Butt joint	168
,, Rising	174
,, Projecting	174

C

	PAGE
Canadian pine	10
Carpenter's bench	22
Carriages for stairs	237
,, Method of forming	244
Casements	201
,, fixed	203
,, hung on centres	203
Cavetto or hollow moulding	178
Cedar wood	13
Ceiling joists	132

	PAGE
Ceilings	221
,, Coffered	222
Centres	89
,, Table of weights on	91
Cesspools	57
Chamfer, sunk	257
,, plain or hollow	257
Chase-mortise	36
Chestnut, its use	2
Chisels	26, 283
Clamping	172
Cleat	77
Close strings	237
Closing style	194
Cogging	35
Collar-beam roof	43
Columns, How to fix	223
,, formulæ for different thicknesses	112
Common rafters	51
,, ,, How to set out	265
Condensation gutter	225
Conical skylight	259
Conservatories	160
Cornice	215
Couple roofs	42
,, close roof	42
Covering to roofs	41
Crane, Derrick	108
Cross garnets	174
Curtail step	236
Cut string	237
Cut and mitred string	237

D

	PAGE
Dados	221
Dancing stairs, Method of laying out	242
Deal, dimensions	21
,, white	1, 9
Decorated mouldings	182
Defects of timber	15
Derrick crane	108
Dogs	81
Domes	248
,, surmounted	248
,, surbased	248
,, Method of setting out	251

	PAGE
Doors	189
,, solid frames	189
,, ledged...	190
,, ledged and braced ...	190
,, framed, ledged, and braced	190
,, framed and panelled...	191
,, how to frame... ...	193
,, how to frame and fix...	277
,, dwarf	194
,, jib	194
,, folding	194
,, double margined ...	195
,, sliding...	195
,, Mediæval	195
,, furniture	198
,, rails	193
,, Renaissance	198
Dormer windows	55
,, in Mansard roof ...	62
Double tenons	186
,, bead and quirk ...	186
Dove-cots	158
Dovetail joints	168
,, ,, lapped ...	170
,, ,, lapped and mitred ...	170
Dragon beam...	55
Drawboring	278
Drips, distance apart... ...	57
Druxiness in timber	15
Dry rot	15

E

Eaves of roofs	55
Echinus moulding	178
Elbow linings	221
Elliptical lantern lights ...	259
Elm, its use	12
Espagnolette bolt	212

F

Fanlight	195
Fascia...	222
Fastenings	278
Felling timber	14
Fender-wall	123

	PAGE
Fillet	176
,, Tilting	225
Fireproof floors	128
Firring to roof	40
,, joists	149
Fliers	236
Flights	237
,, Headroom between ...	240
Flitched beams	114
Floor-boards	135
,, ,, folded	135
,, ,, butt joint ...	137
,, ,, rebated ...	137
,, ,. rebated, grooved and tongued	138
,, ,, ploughed and tongued ...	138
,, ,, grooved and tongued ...	138
,, ,, rebated and filleted ...	138
,, ,, dowelled ...	139
Floors...	121
,, Composite	128
,, Fireproof (Evans & Swain	128
,, Parquet	140
,, Gallery	134
,, specially laid for dancing, &c.	140
,, Weight of	149
Flying shores...	82
Formulæ for finding curvature of bridge	71
Formulæ for breaking weight of beams	114
Formulæ for calculating strength of flitched beams	115
Formulæ for different thicknesses of columns	112
Formulæ for scantlings of joists	123
Formulæ for measuring timber	21
Framed floors	126
Frames, hollow boxed or cased	203
Frames, Louvre, Bevels for ...	272
,, Door	189
Framing	213

		PAGE
Frieze ...		215
Furniture		164

G

			PAGE
Gallery floors...		...	134
Gantry	...	...	103
Gates ...	...	...	160
German oak ...	...	...	10
Gib-and-cottar joint	...		39
Girders, trussed with wood		...	117
,, ,, iron		...	117
,, framed truss		...	119
,, rule for scantlings		...	128
,, dimensions relating to width of bearing			131
Glass windows	...	...	202
Glue ...	...	...	278
Going of a flight of stairs		...	237
Gothic mouldings	...	...	180
Gouge...	...	...	26, 283
Greenheart, its use	...	...	12
Grindstone	...	...	283
Grounds	...	...	214
,, Framed	...	...	215
Guides, Iron, for sliding doors			195
Gumming a saw	...	...	281
Gutters	...	...	55
,, V-shaped	...	...	57
,, Box ...	...	...	57

H

Half-space	..	..	... 237
,, timber work		...	150, 153
Handrail	...	...	... 237
,, Wreathed	...		... 237
,, Iron core to			... 247
Hanging style		...	... 193
Hammer beam		..	... 58
Haunching ...	...	...	... 166
Heart-wood ...	...	...	14
,, -shake ...	...	...	15
Hinge, centre-pin	...		... 172
,, cross garnets...		...	174
,, shutters	...	...	174
,, H and H	...	...	174
,, hook-and-eye		...	174
,, wrought-iron band		...	189

		PAGE
Hinging	... ...	172
Hip rafters	... ...	55, 264
,, roof, Bevels for ...		... 264
Hornbeam wood	...	... 13

I

Inter-tie	... ...	... 164
Iron core to handrail...		... 247

J

Jack-plane	... ...	... 24
,, rafter	... ...	... 264
Jamb linings ...	...	198, 209
Jarrah wood ...	...	... 11
Jib-and-cottar joint	...	... 39
Joggled beams		... 115
Jointer plane ...	...	... 25
Joints, Carpentry	...	30-39
,, Joinery	...	165-172
,, Butt	...	... 168
,, Lap ...	...	31, 170
,, Glued...	...	... 172
,, ,, and blocked		... 172
Joint, Rule ...	...	... 174
., Dovetail	...	... 168
,, Lapped and mitred		... 170
,, Mitre ...	...	... 166
,, Keyed...	...	... 223
,, for beams	...	33
,, ,, halving	...	33
,, ,, notching	...	34
,, ,, cogging	...	35
,, ,, mortise-and-tenon		... 35
,, ,, tusk-tenon...		36
,, ,, chase-mortice		... 36
,, ,, bridle	...	37
,, ,, jib-and-cottar		... 39
Joists, Firring to	...	... 149
,, Ceiling	...	... 132
,, Trimming	...	... 121
,, Formulæ for scantlings of ...	... ...	... 123
,, Scantlings for bridging		125
,, ,, binding		127

K

	PAGE
Keyed joints	223
King-bolt roof	43
,, post	46
,, ,, table of scantlings	50
Knee in staircases	237

L

Landings	237
Lantern	231
,, lights, Elliptical	259
Lap joint	31
Larch wood	13
Latch, Norfolk	198
Laylight	231
Lead aprons	225
Lean-to roof	41
Ledgers to scaffold	100
Linings	189
,, rebated	189
,, jamb	198
,, soffit	198
,, inside	205
,, back	205
,, outside	205
,, elbow	221
,, window	271
Lintels, Wood	206
Louvre frames, Bevelst or	272
,, windows	233
Lychgates	159

M

Mahogany	11
,, its use	11
,, Honduras	11
,, Mexican	11
Mansard roof, showing the setting out	61
Match-boarding	166
Mediæval doors	195
Medullary rays	14
Meeting rails	205
,, style	194
Miter box	29
,, joints	166
,, square	29

	PAGE
Mortise-and-tenon joint 35, 135,	169
,, chase joint	36
Mouldings	176-188
,, *Classic*	176-180
,, fillet	176
,, astragal	176
,, bead and double quirk	176
,, torus	178
,, reeding	178
,, cavetto or hollow	178
,, ovolo	178
,, ogee	178
,, ogee-reversa	178
,, Scotia	178
,, echinus	178
,, raking	259
,, *Gothic*	180-182
,, round bowtell	180
,, pointed bowtell	181
,, roll and fillet	181
,, roll and side fillet	181
,, Decorated	182-185
,, roll and fillet	182
,, roll and triple fillet	182
,, ogee	182
,, double ogee	183
,, scroll	183
,, wave	183
,, plain or hollow chamfer	184
,, sunk chamfer	184
,, *Perpendicular* bowtell	186
,, double ogee	186
,, roll and fillet	187
,, *Modern*	187
,, bolection	188
Muntings	191, 193

N

Nails	278
Needles in shores	76
,, Scantlings of	86
Needle shoring and under-pinning	85
Newel	237
Niches	250
,, Method of setting out	251

T

INDEX.

	PAGE
Norfolk latch	198
Northern pine	9
Nosings	236
Notching of beams	34

O

	PAGE
Oak, Austrian and German	10
,, American	10
,, African	10
,, sill	201
Ogee moulding or cyma-recta	178, 182
Ogee-reverse or cyma-reversa	178
Oilstone	284
Open string	237
Outer string	237
Ovolo moulding	178

P

	PAGE
Packing pieces	225
Palings	162
Parquet floors	140
Parting bead	205
,, slip	205
Partitions	142
,, bricknogged	143
,, quarter	144
,, trussed	146
,, weight of	149
Pendentives	254
Pillars	112
Pine, Northern	1
,, American yellow	1
,, Canadian red	2
,, "Kauri"	2
,, Pitch	2
Pinning to joints	279
Pitching piece	242
Planes	24, 25, 26
Planes, How to use	281
Plank, timber	21
Pocket piece	206
Pointed bowtell	181
Pole plates	51
Poplar wood	13
Porches	159
Princesses	53

	PAGE
Principal posts	150
,, rafters	48
Pulley style	205
Purlins	48
,, Bevels for	269
Putlogs	100

Q

	PAGE
Quarter partitions	144
,, space	237
Queen-post roof	52
,, ,, with princesses	53
,, ,, ,, table of scantlings	54
Quirked bead	168

R

	PAGE
Radial bars	259
Rafters, jack	264
,, hip	55, 264
,, principal	48
,, common	51
,, to set out	264-269
,, valley	55
Rails, top	193
,, frieze	193
,, middle	193
,, lock	193
,, bottom	193
,, meeting	205
Raking moulding	259
,, shores	76
,, scantlings of	79
Ramp	237
Rebates	227
Rebated linings	189
Rebating-plane	25
Reeding	178
Relieving arch	206
Renaissance doors	198
Reveal	206
Rise of stairs	236
Rods, Setting out	273
Roll and fillet moulding	181
,, and side fillet	181
,, and triple fillet	182
,, to apex of roof	231

INDEX.

	PAGE
Roofs	40-62
,, lean-to	41
,, V	42
,, couple	42
,, couple close	42
,, king-bolt	43
,, collar-beam	43
, scantlings for	45
,, king-post truss	46
,, hammer-beam	58
,, Mansard	61
,, hip	264
,, coverings to	41
,, eaves to	55
,, firring to	40
,, valley, the setting out	268
Rot, dry	15
,, wet	15
Rough string or carriages	237
Round bowtell moulding	180

S

	PAGE
Sapwood	14
Sashes	201
,, circular headed, in circular wall	258
,, bars, vertical	202
,, ,, horizontal	202
,, lines	205
,, bars, capping to	227
Saws	22, 23, 280
,, gumming	281
,, setting	280
,, filing	280
Scaffolding	99
,, braces for	101
Scantlings of girders, rule for	128
,, for bridging joists	125
,, for binding joists	127
,, king-post, table of	50
., needles	98
,, queen-post roof	54
,, collar-beam roof	43
,, raking shores	79
,, couple close roof	43
Scarfing, proportions for	32
., to resist tension	32
,, ,, compression	32

	PAGE
Scarfing, to resist cross strains	32
,, ,, tension and compression	33
Scotia moulding	178
Screws	279
Scroll moulding	183
Seasoning of timber	16
Shoes	279
Shop fronts	222
,, ventilation	222
Shores	76-88
,, raking	76
,, scantlings of	79
Shores, angle of sole piece	81
,, forms of dogs	81
,, flying	82
,, calculation of thrust	83
,, needle	85
Shutters, folding	209
,, lifting	211
,, external	212
,, hinge	174
Sill oak	201
Skylights	225
,, conical, on an elliptical base	259
Sleepers	106
Slip feathers	166
Smoothing-plane	25
Socket piece	279
Soffit linings	198
Sole piece	77, 78
Space, quarter	237
,, half	237
Spandril framing	239
Spikes	279
Spokeshave	27
Square, the	27
Square of timber	21
Stairs, straight	239
,, dog-legged or newel	239
,, open newel	239
,, circular newel	239
,, circular geometrical	239
,, dancing, method of setting out	242
Staircases	236
,, on planning	240
,, geometrical	244

	PAGE		PAGE
Stall-board	222	Tools	22-29
Standards	99	,, saws	22
Star-shake	15	,, planes	24
Step, curtail	236	,, chisels	26
Strain, weight per ft. super. to		,, brace and bits	27
allow for on bridges	75	,, various	28
Stirrups	128, 279	,, the use and sharpening	279
Straps	279	,, how to preserve from	
String, wall	237	rust	284
,, outer	237	Torus moulding	178
,, cut	237	Transoms	195
,, cut and mitred	237	Traveller	106
,, open	237	Tread	236
,, close	237	Trenails	278
,, rough, or carriages	237	Trimmer	121
Struts	48	,, arch	132
Stuff	213	Trimming joists	121
Styles, hanging	193	Trussed girder bridge	68
,, closing	194	Trying plane	24
,, meeting	194	Turrets	157
Suspending pieces	48		
Suspension bridge	70	**U**	
		Underpinning	87
T		**V**	
Tables of least rise for different spans of curved rib bridges	65	V-roof	42
		V-jointed boarding	166
Tables of weights for centres	91	Valley rafter	55
,, of scantlings for bridging joists	125	,, ,, How to set out	268
		,, roof, Bevels for	267
Teak, characteristics of	11	Veneer	224
Tenon	35	Ventilation, shop	222
,, double	186	,, round timber	133
,, tusk	36		
Tie-beam	48	**W**	
Tilting fillet	225		
Timber, half timber work	153	Wall, Fender	123
,, load of	21	,, piece	76
,, its growth	14	,, plates	132
,, felling	15	,, string	237
,, defects	15	Walnut, its use	13
,, good qualities	16	Water-bar	201
,, seasoning	16	Wave moulding	183
,, conversion of	18	Weep-holes	231
,, formula for measuring	21	Weight of flooring	149
		Weights, Window	205
,, market form	21	Well-hole	239
,, definitions of	21	Wet rot	15

INDEX.

	PAGE
White deal	9
Wind force, allowance in roofs	51
Winders in stairs	236
Window board	208
,, blinds	209
,, weights	205
,, furniture	212
,, backs	221
,, linings, splayed bevels for	271
,, glass	202
Windows	201
,, dormer	55
,, louvre	233
Wood, various	13
,, block floors	139
,, defects in	15
,, sill	150
,, lintels	206
,, bricks	206
Woods, Weight of—	
Northern pine 32lb. per c.ft.	9

	PAGE
Woods, Weight of—	
American yellow pine 35lb. per c.ft.	9
Australian pine, "Kauri," 36lb. per c.ft.	10
Oak 48lb. per c.ft.	10
African oak 57lb. per c.ft.	10
Mahogany 64lb. per c.ft.	11
Teak 47lb per c.ft.	11
Greenheart 70lb. per c.ft.	12
Ash 50lb per c.ft.	12
Elm 42lb. per c.ft.	12
Walnut 41 to 43lb per c.ft.	13
Beech 45lb. per c.ft.	13

Y

Yarrah wood, its use	11
Yellow deal	9
,, pine	9

www.ingramcontent.com/pod-product-compliance
Lightning Source LLC
Chambersburg PA
CBHW032045230426
43672CB00009B/1481